Digital Mind Math

Robert Paster

ISBN: 1532775946

ISBN 13: 9781532775949

Library of Congress Control Number: 2016906511

Contents

Introduction

There are only two complete numbering systems. One complete numbering system—a numbering system that allows us to form a completely filled-in number line—is the familiar set of real numbers that we learn about in school. The other complete numbering system—called the p-adic numbers—is the numbering system that creates the mathematics of thinking, the natural mathematics of cognition, the mathematics of the mind.

With this brief, three-sentence introductory paragraph, we can see the whole book ahead of us: What makes a numbering system complete? How can there be only two complete systems of numbers? What's wrong with real mathematics as a model of how we think? What is p-adic mathematics? How is p-adic mathematics the mathematics of the mind?

That's our agenda. Let's begin.

A Different Mathematics

This book, *Digital Mind Math*, describes the human way of thinking, which is based not on our usual real mathematics, but instead on the only other complete numbering system, the p-adic numbers, the numbers of the natural mathematics of cognition.

We should probably immediately digress to the statement that there are only two complete numbering systems, real and p-adic. This will be discussed in upcoming chapters, but for now the discussion will be introduced with the clarification that these are the only two ways to allow representation of all numbers, to fill in all the holes between numbers. Except for these two number systems, all other numerical systems are incomplete.

This sounds pretty unbelievable, but—assuming it's true—it also sounds pretty important. We'll postpone discussion of how this was proven, but for now let's just note that this gives a special role to p-adic mathematics: There are two and only two complete numerical systems, capable of representing the set of all possible numbers. The mind works using one of these number systems, p-adic numbers. P-adic mathematics is how we think.

P-adic mathematics is our first mathematics. It is how we become familiar with the world, how we structure the world and our place in it, how we learn, how we know, how we remember. This will be discussed in more detail in upcoming chapters.

But there is one more aspect of how p-adic mathematics achieves the mind's great cognitive efficiency. This second aspect has to do with a reconfiguration of spacetime—a reimagined structuring of space and time—a reconfiguration that fits p-adic mathematics like a glove.

Reconfigured Spacetime

Spacetime is a term of modern physics. Digital Mind Math—the language of the mind, as detailed in this book—is part of a theory of modern physics. This will be discussed in upcoming chapters.

For the moment, we begin by moving away from thinking of a person, or a piece of furniture, taking up a section of preexisting three-dimensional space. Instead we think of the person or the piece of furniture each as its own universe, from its moment of inception to its moment of demise, extending in three dimensions of space, staying the same or changing as it passes through the fourth dimension, time.

Our mind conceptualizes these four-dimenional universes—people, chairs, tables, rooms, buildings—and elaborately interconnects them in combinations and permutations to create thought fragments, which we are adept at recalling and revising and projecting into the future in varied scenarios. P-adic mathematics is built to do this, with a level of ease and efficiency that real mathematics cannot match, because p-adic mathematics is the mathematics of nestings of enclosures.

In this book, we will show how our mind operates by manipulating these intricately interconnected nestings of enclosed universes.

Quantum Physics

Explaining cognition as the application of p-adic mathematics to this reconfigured spacetime is part of a theory of modern physics, developed by Finnish physicist Matti Pitkänen. This theory, which is called Topological Geometrodynamics, or TGD, uses the same processes to explain particle physics, cosmology, biophysics, and the mind.

It is a book-length discussion, published as the author's previous work, *New Physics and the Mind*, to present the argument in favor of Pitkänen's TGD as the most important radical theory of modern physics. For the moment, we note that TGD's core cognitive process is the same as the core quantum process that, in generally accepted modern physics, takes place at the level of elementary particles.

This core quantum process, an entirely standard aspect of quantum physics at the level of elementary particles, is the quantum jump, also referred to as the quantum transition, or the quantum leap, or quantum decoherence, or the reduction or the collapse of the quantum wave function. In accordance with this process, an elementary particle (an electron, for example) proceeds in a configuration space, along every possible path, until it collapses momentarily at a single actualization upon observation or interaction or measurement. At the quantum jump, the electron momentarily transitions from the sum over histories—quantum physics' accumulation of all possible paths—to a single classical event.

In TGD as well as in Digital Mind Math, this core quantum process—the quantum jump—is also the model of how we think:

generating a set of possible next thoughts, then tending to select, as the thought-line to proceed along, that thought-line which maximizes information content.

Maximization of information content is a topic that we discuss at great length in the chapters ahead. Perhaps surprisingly, this is a well-defined mathematical concept. It is also a well-defined concept of physics: the opposite of physics' entropy.

Entropy is disorder or randomness, lack of information. So negative entropy—negentropy—is information. The mind is guided by the principle of negentropy maximization, or maximization of information content.

The mind—using its natural language, p-adic mathematics—generates possibilities for our next thought, and, guided by the negentropy maximization principle (the maximization of information content), selects the next thought to experience.

This real thought that we experience—momentarily, before circling back for another round of possibility generation and analysis—is experienced as a real phenomenon, the real mathematical analog to the selected p-adic thought.

The processes of possibilty generation and selection are most efficiently accomplished using the mind's natural language, p-adic mathematics. This then triggers the real experience of the selected thought—the set of mind images and body experiences that our mind has labeled p-adically.

We will discuss the concept of dual labeling—p-adic mathematics for analysis, real mathematics for experience. In the Afterword, Dr. Pitkänen will present some of his own more detailed thoughts about the relationship between p-adic labeling and real labeling, and will also present his thoughts on the subject of where it is that thoughts and memories reside.

Summary So Far

P-adic mathematics, applied to a physics-based reconfiguration of space and time, gives us a simple way to record and analyze our thoughts. The thoughts that we experience are the mathematically real analogs of the p-adically labeled thoughts that have been analyzed then selected in order to maximize information content.

Just One Outrageous Claim

When you think about it, it's pretty outrageous to claim that there is a mathematics of the mind, that we can digitally model how we think. That is this book's outrageous claim. But it is the only outrageous claim that *Digital Mind Math* will focus on.

I say this because I don't want you jumping ahead to think that this book is also answering a host of other related questions about the mind or the brain or computers or artificial intelligence, besides the question of mathematically modeling how we think.

One outrageous claim is enough, and it will take this whole book to flesh out the details of the claimed mathematical model of the mind.

I feel obligated to rein you in a bit, even before we begin, because, if one is convinced that there may be something to p-adic mathematics as the mathematics of the mind, it's tempting to leap ahead to all kinds of other questions that this book will not be answering.

My mission will be accomplished if you understand why p-adic mathematics, applied to reconfigured spacetime, gives us the mathematics of how we think. This will take the whole book to do. Then you may feel free to branch off into additional areas of your own expertise to see how Digital Mind Math applies.

For example, this book will not answer the deep questions

about the philosophy of mind. It will not say whether our mind is a soulless automaton or a divine inspiration. It will not say whether there is free will. Generally, p-adic mathematics appears to accommodate any reasonable answer to these deep questions. There will be points in *Digital Mind Math* when we will comment briefly on some of the philosophical issues related to the mind. But p-adic mathematics itself is a tool, a complicated tool, but one that this book will use only to model how we think.

Nor will this book explain the neurobiology of the brain. As you become more comfortable with p-adic mathematics' Digital Mind Math, it will be very tempting to see how well this could model the workings of the brain and its neurons and synapses. In *Digital Mind Math*, we will occasionally comment when the description seems to be that of a mind that could fit right onto the brain, an exact mapping without the need for any intervening processes or mechanisms. However, I will leave to neuroscientists and other experts on the brain the determination of the details of this fit and of how precise this fit is.

Nor is this a book about modern physics. Although this book's Digital Mind Math is the model of the mind that is part of the radical theory of physics TGD (Topological Geometrodynamics), TGD's scope is much broader. TGD is a theory of particle physics and a theory of cosmology, as well as a theory of biophysics and a theory of the mind. It is only the aspect of TGD's theory of the mind that is addressed in this book. For particle physics or the physics of the universe or biophysics, the interested reader may go directly to the published writings of TGD's developer, Finnish physicist Matti Pitkänen, with thousands of pages of Dr. Pitkänen's writings easily accessible online. As of this writing, the best source for this is Dr. Pitkänen's TGD website www.tgdtheory.fi. Dr. Pitkänen also regularly publishes musings on physics and consciousness at www.matpitka.blogspot.com.

Nor is this a book of postdoctoral mathematics. This book describes the basics of p-adic mathematics. But the full scope of p-adic mathematics is extraordinarily advanced, a specialized subject of advanced academic mathematics. P-adic mathematics is our natural mathematics of thinking and our first mathematics, but it will take this whole book to explain how simple it is.

Nor is this a book answering all of the important design questions of artificial intelligence or computer science. Again, these applications—like applications to neuroscience and the organization of neurons and their connections—are tempting to hypothesize about, and we will on occasion succumb to this temptation. But *Digital Mind Math* is focused specifically on how the mind works, rather than how computers work or might work. It may be obvious to some that the best approach to artificial intelligence is to follow the path of human intelligence. But first things first: Let's see if we can develop a common understanding of the mind.

A Note About Terminology

There is no uniform way in which the "p" in p-adic mathematics is written in the mathematical and scientific literature. However, most commonly, the "p" is italicized, so that this branch of mathematics is written as *p*-adic mathematics.

This lower-case italicization is often (although not always) extended to the beginning of a sentence; for example: *p*-Adic mathematics is my favorite form of mathematics. The rationale behind this notation is that p stands for prime numbers, and the italicization emphasizes that it's not just an abbreviation, it's a representation of a specific (although variable) prime number.

This book does not adopt the lower-case italicized p. And if the first word of a sentence is p-adic, it will be the p, not the a, that will be capitalized. The reason for this is an attempt to demystify this mystifying mathematics, to make p-adic mathematics seem like a more natural, normal term, not one that raises a question (why is that p italicized and not capitalized?) even before you've become comfortable with it. Also, a more normal spelling (not italicized; capitalized when it's the first word of a sentence) seems more compatible with the short messaging and simplified rules such as we see in today's popular messaging formats.

Another Usage Note

This book has a deliberately casual or conversational tone. The purpose of this is to try to put the reader at ease in an attempt to make the mathematical and scientific information presented a bit more accessible. Believe it or not, some people get nervous when they're told that the subject being discussed is quantum physics or higher mathematics. So it's worth trying to put readers at ease.

There is also the issue of using the word "they" as a gender-neutral singular pronoun, instead of the awkward "he or she," or the sexist "he." You will recall that "they" used this way was selected by the American Dialect Society as the 2015 Word of the Year.[1] Ahead in *Digital Mind Math*, we will at times randomly select "she" or "he" when gender isn't known, and at other times we will use the neologistic gender-neutral singular "they."

Another grammatical quirk that we'll find ourselves using is the use of "we" or "our" when we really mean "I" or "my," or "you" or "your," or "he or she" or "his or her." For example, we'll talk a lot about "our mind" when maybe more precisely we should be saying "my mind," or "your mind," or "his or her mind," or (plural) "our minds," "your minds," "their minds." In part, this isn't a neologism at all, but rather the old elementary school teacher trick of talking about what "we" should be doing when what is really meant is what "you" or "he" or "she" should be doing. I apologize in advance if you feel infantalized by this usage as a misguided attempt at being casual. But the use of the singular "mind" in "our mind" is neologistic, similar to the gender-neutral singular "they," but in this case not relating to gender, but instead an attempt to meld the perspective and mind of the reader with that of the writer.

What's Upcoming

Part One of this book will explain the basics of Digital Mind Math, enabling readers to see how p-adic mathematics, applied to reconfigured spacetime, is exactly how their everyday, moment-by-moment thinking proceeds.

Part Two offers some mathematical detail.

Part Three then applies Digital Mind Math to numerous examples, strengthening the reader's intuition for how Digital Mind Math works.

Part Four belies the earlier statement—that *Digital Mind Math* will be making just one outrageous claim—by making a second outrageous claim. Maybe we can consider Part Four's second claim as Claim Number 1A rather than Claim Number 2. In Part Four, we expand on the earlier claim—that we can mathematically model how we think—by proposing that there is a correct way to think. Part Four expands on an aspect of p-adic mathematics to propose that there is a natural, mathematically defined specification that optimally takes full advantage of all the organizational capability that p-adic mathematics offers.

After a brief Conclusion, the reader will have the pleasure of reading an Afterword written by Dr. Matti Pitkänen, the developer of Topological Geometrodynamics (TGD), the extraordinary theory of modern physics upon which Digital Mind Math is based.

An Appendix is also included with suggestions for theoretical Digital Mind Math issues that experts in p-adic mathematics can resolve in order to push forward the development of a workable model of Digital Mind Math as the basis, for example, of a revised approach to artificial intelligence.

PART ONE: THE BASICS

The elementary concepts of Digital Mind Math are presented in Part One. Each concept is fairly ordinary, in that it may have crossed the mind in idle speculation about life and the universe. Taken together, though, these basics provide a revised worldview and a model of how we think—the mathematics of the mind.

Digital Mind Math is based on the theory of physics Topological Geometrodynamics (TGD). So this whole book, *Digital Mind Math*, is about TGD. The scope of TGD is much larger than the scope of Digital Mind Math, however, since TGD is a theory of modern physics, and Digital Mind Math is a detailed development of the aspect of TGD having to do with thinking and the mind.

Parts One, Two, and Three of *Digital Mind Math* will each be introduced by a summary of its chapters. Readers may feel free to take advantage of these summaries by reading chapters in a different order or skipping chapters for which the brief summary seems sufficient.

Chapter 1. Enclosings. Digital Mind Math emphasizes the enclosings of the physical world. Objects at every level enclose and are enclosed by other objects that each are virtual universes— bounded universes, finitely extended, ending at their outer boundaries. In Topological Geometrodynamics, and therefore in Digital Mind Math, an object is represented by a bounded spacetime sheet, enclosed within and enclosing multiple levels of other spacetime sheets. This is many-sheeted spacetime, a hierarchy of bounded universes.

Chapter 2. Connecting Enclosures. Spacetime sheets are connected through tiny wormholes. These tiny wormholes implement the organization of many-sheeted spacetime's enclosings. Wormholes are the connections from one macroscopic and macrotemporal four-dimensional spacetime sheet to another. Mathematicians are able to model these four-dimensional spacetime sheets and their four-dimensional wormholes very simply, in eight dimensions. TGD's and Digital Mind Math's eight-dimensional spacetime consists of four dimensions of macro-extended spacetime—four dimensions of spacetime extended macroscopically and macrotemporally—and four tiny dimensions of space in which the connecting wormholes reside.

Chapter 3. Thinking Enclosures and P-Adic Math. Thinking proceeds by processes of enclosing and becoming enclosed. Digital Mind Math uses a special mathematical system, called p-adic mathematics, to model both the process of thinking and the organization of real spacetime sheets within many-sheeted spacetime. The basics of p-adic mathematics are introduced.

Chapter 4. Our Blinking Moments of Time. Spacetime sheets extend not just in three spatial dimensions, but also in the fourth dimension—time. As a result, an object in TGD—a four-dimensional spacetime sheet—is a time history of the object's extension in three spatial dimensions. Wormholes, connecting TGD's spacetime sheets, reside in four tiny dimensions of space. Our mind advances from moment to moment according to the processes of physics' quantum jump. The mind applies p-adic mathematics to label the spacetime sheets within TGD's eight-dimensional spacetime, which consists of four-dimensional spacetime sheets hierarchically linked by four-dimensional wormholes.

Chapter 5. Review of the Basics. The real world is most effectively understood as embedded and embedding objects proceeding along a configuration space of all possible paths—a world of classical worlds—until reduction to a single path when observation or measurement triggers a quantum jump. The organization of the embeddings is most naturally tracked not by real mathematics but by p-adic mathematics, the mathematics of enclosure. P-adic mathematics is the natural mathematics of thinking, memory, imagination, and intention. Our biological systems demonstrate both real and p-adic properties that permit mind-body integration.

Chapter 6. Memories. Memories are recorded within the p-adic structure of the brain. When a memory is triggered, the corresponding real structure directs our experience.

Chapter 7. The P-Adic Surgery of Dreams. Dreams are discussed as everyday phenomena that exhibit the mind's use of p-adic and real mathematics, and that, guided by information maximization, select life paths from a configuration space of possibility.

Chapter 8. Seconds, Minutes, Hours, Days, Weeks, Months, Years. We use several mathematical approaches in thinking about elements of time—purely human-invented numerical relationships, astronomical relationships based on the moon, astronomical relationships based on the sun. This chapter discusses the interactions among these mathematical relationships, as practice for thinking about the relationship between real and p-adic mathematics.

Chapter 9. Cognition, as Modeled by Piaget and Vygotsky. Jean Piaget's and Lev Vygotsky's twentieth-century models of cognitive development remain core to contemporary psychological and educational theory. Piaget's framework—assimilation, accommodation, equilibration—is precisely modeled by TGD p-adic mathematics. And both theories give us valuable insight into the drive to increase information content.

Chapter 10. Computers vs. the Mind. For many years now, computers have played chess better than the world chess champion. Computers beat the champions of television game shows and the board game Go, and show remarkable agility at tasks such as facial recognition that only a few decades ago seemed humans would always be superior at. But the human brain weighs less than three pounds and uses the energy of a dim light bulb. P-adic mathematics, applied to reconfigured spacetime, accounts for humans' greatly superior efficiency.

1. Enclosings

One of the basic new ways that TGD asks that we structure reality is to look at the world as many-sheeted encloseds and enclosings.

The bird on the tree across the street encloses a set of organs, which each enclose a set of cells, which each enclose a set of organelles (cell parts), which each enclose a set of molecules, which each enclose a set of atoms, which each enclose a set of elementary particles. The bird is enclosed by our Earth system, which is enclosed by our solar system, which is enclosed by our galaxy, which is enclosed by the universe.

TGD asks you to conceptualize the world as sets of encloseds and enclosings.

And TGD asks you to consider each object as a universe, a bounded universe, finitely extended, ending at the object's outer boundary.

Some Introductory P-Adic Concepts

It's probably natural to think in real images about the bird, and all the layers that the bird encloses, and all the layers that enclose the bird. We'll be referring to this layering as "many-sheeted spacetime," which is realized both in the real sense that these images images evoke, as well as in a p-adic sense.

P-adic numbering abstracts and simplifies, dropping away irrelevant details, and recording the pattern of enclosures. The real spacetime sheets have p-adic geometric correlates that are labeled p-adically and represent thought bubbles—basic units of cognition.

Two people do not necessarily conceptualize the levels of enclosure the same way. How we conceptualize the levels—the bird enclosed by an Earth system, or by a local climatic system, or by a geographic area, or by a city or a country or a province, or by a branch of a tree—is individually and flexibly defined, based on each individual's past history of thought and knowledge and observation, and even flexibly accessible as different conceptualized enclosings based on the individual context. When fully realized, p-adic mathematics permits intermediate levels to be introduced or ignored without any disruptive renumbering or reconceptualization.

We'll be detailing this, conceptually and mathematically, in upcoming chapters. For now, we focus on how this p-adic TGD way of looking differs from how we normally think about looking.

The Normal Way of Looking

Normally, we think of a physical object as occupying a section of three-dimensional space. We think of space as continuing on in one dimension infinitely to the left and infinitely to the right, and in a second dimension infinitely forward and infinitely back, and in a third dimension infinitely up and infinitely down. Within this structure of space, we consider an object—say that bird on a tree—to be occupying a section of space.

We think of the bird as an object that is occupying a finite, bounded section of infinitely extended space.

This is not how TGD views the bird. So, since we plan to retune our minds and senses so that we think and live TGD, we'll want to start looking at that bird the TGD way.

The TGD Way of Looking

In TGD, the bird—any object—is a finitely extended universe, ending at its outer boundaries.

We're aware of the organs, cells, organelles, and so on that the bird encloses, and we note the various hierarchies of enclosure.

In TGD, this hierarchical structure is the structure of both the real spacetime sheets and the p-adic spacetime sheets.

In TGD, many-sheeted spacetime is real—the real structure of the universe and everything in it. And the p-adic spacetime sheets provide cognitive representations that correlate to thoughts and permit imagination.

For thoughts, tracking the levels of enclosure p-adically does not require us to track the relative sizes of the various enclosed and enclosing objects. We need only track the hierarchy level of what's enclosing what.

The p-adic perspective of enclosures is ultrametric: This perspective is indifferent to measurement, attending only to hierarchical level of enclosure.

Our cognitive perspective of enclosures is non-Archimedean, not measured. So this perspective will have a different arithmetic than our normal real arithmetic, and will have a different geometry than our normal real geometry.

TGD requires us to be aware of both the real metric perspective, arithmetic, and geometry, and the ultrametric perspective, arithmetic, and geometry of enclosure. The nature of this ultrametric mathematics, and how this ultrametric mathematics relates to real mathematics, are important aspects of TGD, which we will be exploring in the upcoming chapters.

Not a Completely Novel Thought

This perspective of enclosure is not at all a completely novel idea, one that Pitkänen has invented from scratch, one that has never been contemplated. On a number of occasions—moments of conversational speculation, and moments of introspection—I for one have crossed paths with this idea.

For example, has it ever crossed your mind that that liver of yours may have its own existence, its own circumscribed reality, in addition to its reality as part of your body? And isn't there a wholeness to your arms, and to your fingers, and to your eyes? Don't these body parts seem to exist as their own finitely extended universes, their own sheets within a many-sheeted hierarchy?

This is not a radical picture of reality, this many-sheeted reality. But Pikänen is asking us to take this many-sheeted reality very seriously, as a basic physical organizing principle, core to biology, cosmology, and physics.

This is our first basic principle of living TGD.

2.   Connecting Enclosures

In TGD, a bird is its own four-dimensional universe, a four-dimensional spacetime sheet. The boundaries of this four-dimensional bird-universe are its extension in its three spatial dimensions, as it varies over the length of its fourth dimension, time.

The bird encloses its organs, but each organ is its own universe, its own four-dimensional spacetime sheet.

This forces a very different structure of the TGD physical world than in traditional (non-TGD) physics. In traditional physics, the bird occupies space in a section of universal space's three spatial dimensions. The bird's left eye occupies part of the space that the bird occupies.

But in TGD, the bird is a universe, and the bird's left eye is another universe. The bird occupies its entire spacetime sheet—it's all bird, all a bird-universe.

The eye is another universe and does not occupy the same spacetime sheet as the bird. Instead, the eye—in its own universe, its own spacetime sheet—is connected to the bird by a small wormhole. The bird-universe is a physically larger four-dimensional spacetime sheet (that is, when measured, using real mathematics, the bird is metrically larger than its eye). A wormhole within the bird-universe connects it to its left eye, a physically smaller four-dimensional spacetime sheet. Within the bird-universe, the only presence of its left eye is as a miniscule wormhole opening. But this wormhole—miniscule within the four-dimensional bird-universe—opens up at the other end of the wormhole to the entire universe of the bird's left eye.

Too Much for Four Dimensions

This four-dimensional bird-universe and four-dimensional eye-universe seem to be too much to squeeze into only a total of four dimensions: The bird spacetime sheet includes the entire expanse of the bird, without the details of the inner workings of its eye, which is its own universe, its own four-dimensional spacetime sheet. If this is the structure, four dimensions aren't going to be big enough for both the bird and its eye.

In addition, we have to deal with the wormhole connecting the bird with its eye. Although the wormhole is very small—each dimension is roughly 10^{-30} meters— it does nevertheless have a spatial extension in four dimensions.

TGD seems to have much too much going on to squeeze these embedded spacetime sheets and their wormhole connections all into just traditional (non-TGD) physics' four dimensions.

TGD solves this problem by structuring the universe as eight-dimensional.

In TGD, there are four macro-extended dimensions of spacetime that are embedded within eight dimensional embedding space. This is best pictured by replacing each point of macro-extended four-dimensional spacetime, which we label M^4_+ spacetime, with a minuscule (10^{-30} meter) four-dimensional space, which we label as CP_2 space

In eight-dimensional embedding space, instead of four-dimensional spacetime M^4_+ consisting of a bunch of zero-dimensional points, M^4_+ consists of a bunch of miniscule four-dimensional CP_2 structures.

In TGD, four-dimensional spacetime sheets reside within eight-dimensional embedding space. Wormholes, which connect the macro-extended spacetime sheets, reside in the four miniscule dimensions. The macroscopic and macrotemporal four-dimensional sheets of many-sheeted spacetime reside in the four extended dimensions of spacetime.

There's a certain intuitive appeal to the small wormholes

being in their own small four dimensions. But it may be harder to accept that all of the infinite layers of the four-dimensional macro-extended spacetime sheets appear in just four dimensions. It seems to require four dimensions for each macro-extended spacetime sheet, so how can we get away with just four dimensions for the whole infinite hierarchy of embedded and embedding macro-extended spacetime sheets?

The answer to this lies in the higher-dimensional structure (higher than four dimensions) of TGD spacetime: This is how we get enough room for an infinite layering of sheets of four-dimensional spacetime. This is why it is sufficient to have eight dimensions, and not to require an infinite number of dimensions, in order to allow an infinite hierarchy of four-dimensional spacetime sheets.

It is the physics of TGD—which must conform to observed physical measurements—that dictates that TGD embedding space is eight-dimensional. But actually, any number of higher dimensions—even, say, just a fifth dimension—would make possible the infinite hierarchy of many-sheeted spacetime.

Let's see why.

How Higher-Dimensional Spacetime Makes Room for an Infinite Hierarchy

We don't need an infinite number of dimensions to allow for an infinite hierarchy of four-dimensional spacetime sheets. We just need the four-dimensional spacetime to be embedded in higher-dimensional spacetime. To see this, we'll illustrate with a one-dimensional world, and show how expansive this world gets if we add an extra dimension.

Now a one-dimensional world doesn't have to be quite as plain as you might think: The dimension does not have to extend along a straight line. In fact, this one dimension could spin and twirl every whichway—up, down, and sideways—as long as we understand that this one-dimensional universe exists only along this squiggly line.

Suppose George is a one-dimensional living object, finitely bounded, lying along a segment of our one-dimensional world's squiggly line. Remember that, in TGD, George's eye is in a different one-dimensional spacetime sheet than George. George takes up all of his finitely bounded space, and his eye is connected to his space but is not part of his space. How does a small second dimension give us room for George and his eye? And, as a greater challenge, how does this same small second dimension also give us room for the eye's iris and cornea and other structural components, and for each component's molecules, and for each molecule's atoms, and for each atom's elementary particles? How does just one added dimension allow room for all of these sheets of many-sheeted spacetime (in the case of our simplified example, room for all of these sheets of many-sheeted one-dimensional space)?

The answer lies in the second dimension, no matter how small, that extends from one-dimensional George outward within a narrow ribbon of two-dimensional surface. George is still one-dimensional, his own squiggly line segment. But George's eye does not lie along that squiggly line segment. George's eye is connected to George by a tiny wormhole that extends outward—within the narrow two-dimensional surface—from the squiggly line segment that is George. George's eye is another squiggly line, within a wormhole length's separation from George, within our two-dimensional world's narrow ribbon of two-dimensional surface. George's eye is its own universe, hierarchically connected to George but on a different squiggly line segment, separated within a wormhole's length from George.

And George's eye's iris? It too is its own universe. The iris is connected to the eye by a second wormhole within our two-dimensional world's narrow two-dimensional surface, a wormhole hierarchically connecting the eye segment with the iris segment.

George exists in one dimension. George's eye also exists in one dimension, not as a segment of George, but as a segment elsewhere in our two-dimensional world, hierarchically linked to George by a wormhole residing along the surface of the narrow second dimension.

The aggregation of all the sheets of George's many-sheeted spacetime exists in two dimensions, and it takes just a tiny second dimension to permit the full embedded hierarchy of the bounded one-dimensional universe George segment and all the other bounded one-dimensional universes that he hierarchically embeds and is embedded by.

Intricately Interconnected

So embedding spacetime sheets within a higher-dimensional embedding space—for George, this meant adding one microscopically extended dimension to his one macroscopically extended dimension—permits the infinite hierarchy of TGD's many-sheeted spacetime. The connecting wormholes reside within the higher dimensions, and permit intricate interconnection among the macro-extended spacetime sheets.

The four microscopic dimensions of CP_2 space, though very small, are an integral aspect of TGD physics, explaining, for example, what otherwise seem to be purely arbitrary constants assumed in the standard model of quantum physics. But these four dimensions of CP_2 space, and the wormholes that reside there, are smaller than the resolution at which we experience the world—too small to see. We therefore experience the world as four-dimensional. We are left with the impression that we live in a four-dimensional world.

CP_2 space being four-dimensional offers a critical advantage to Digital Mind Math: These four dimensions permit unlimited interconnections among the M^4_+ sheets of macro-extended many-sheeted spacetime. With four-dimensional wormholes, the macro-extended M^4_+ sheets can be intricately interconnected, not just one spacetime sheet uniquely connected to one other spacetime sheet. These intricate interconnections that four-dimensional wormholes permit—each thought bubble in relationship to many other thought bubbles—are an important feature of Digital Mind Math.

We'll continue our efforts at visualizing TGD's eight-dimensional spacetime in upcoming chapters.

A brief comment on notation. In the mathematical literature, the complex projective space CP_2, which we're writing with the 2 as a subscript, often appears as CP^2, with the 2 as a superscript. In *Digital Mind Math*, we'll use the subscript notation, as it appears throughout Dr. Pitkänen's documentation of Topological Geometrodynamics.

3. Thinking Enclosures and P-Adic Math

A number of theorists have concluded that thinking proceeds by processes of enclosures.

Let's examine why.

A Day in the Life

Today I have to go to the grocery store.

(Enclosed thought:) Milk. Juice. Bananas. Paper towels.

(Up two levels, through the enclosing grocery store thought, to an errand thought which encloses the grocery store thought:) I'll go to the post office first, then get gas, then go to the grocery store.

(Through a wormhole linked to the errand thought:) Maybe I should take a side trip to look at some shoes.

(Enclosed thought:) These scuffed up shoes are starting to look a little embarrassing.

(Up two levels, through the enclosing shoes thought, back to a now-expanded enclosing errand thought:) Sounds like a plan: Shoes. Post office. Gas. Groceries.

What's Going On Here?

As we bounce around in our minds, we add details, then bring those details into a different context, then detail that different context, then find another detail that takes us somewhere else, then expand on that and go elsewhere, then . . .

Astoundingly, there is a system of mathematics that works this way. A system of mathematics that identifies encloseds and enclosings, that bounces from a large thought to a detail, then finds that detail in another enclosing thought.

This system of mathematics tracks the hierarchies of objects' enclosures, and is the system of mathematics that in TGD is the natural mathematics of thinking, of cognition, of imagination, of intention.

This system of mathematics is called p-adic mathematics. In TGD, p-adic mathematics is the natural mathematics of the mind's thinking processes, as well as the mathematics that naturally tracks the hierarchies of enclosures of real objects in many-sheeted spacetime.

P-adic mathematics and its intertwining with real mathematics are concepts at the heart of TGD. These concepts lead to TGD physics that is intimately connected with consciousness and the mind. And they lead to TGD biosystems built with capabilities to sense and organize in both real and p-adic terms, and to mediate between the real and cognitive worlds, cognizing reality and translating intention to action.

P-Adicizing the Errands

Let's preview the general structure of Digital Mind Math.

Our life experience to this point, as it is placed in the context of all past and possible future experiences, has led us to the conclusion that it's time to go to the grocery store.

After experiencing this thought, our mind's analysis, tending toward increased information content, leads us to detail what we need to buy at the grocery store. These details—this grocery list—change our focus from the single event of going to the grocery store to the components that are enclosed within that event. To maximize information content, after the thought about going to the grocery store, we decide that the next thought ought to be to list what groceries to buy.

Perhaps we've made this easy for ourselves by accessing a preexisting shopping list, as a single event. Or perhaps our mind has to go through a number of information-maximizing steps which link the event of going to the grocery store to the list of what to buy—

checking the refrigerator and the cupboards, thinking about who's coming to dinner, assessing our budget restrictions, and so on.

The grocery list accomplished, we have other information-maximizing decisions to make: We move up two levels of enclosure—from the grocery list, to the grocery store experience, to the whole errand experience that the grocery store is part of—and tend toward maximizing the information content of the whole errand experience.

Here you may start to feel that information maximization is too limiting a concept, that it's forcing a narrow label on a broader experience. This is a valid concern, and one we'll postpone addressing. For now, allow yourself a broad definition of information maximization, as a way of allowing insight into the scheme of enclosings and encloseds.

Our mind has now tested out many scenarios, scenarios that derive from our life experience to date, and our mind has now directed us not toward getting into the car and heading to the proposed list of errands, but rather toward the possibility of buying shoes.

The intricately interconnected possibilities for future events, influenced by the accumulation of events which we have experienced to date, have led us from the errand concept through a wormhole that leads out from the concept of errands.

During our mind's process of generating possible next thoughts in follow-up to the thought about our errand itinerary, our mind has, through a wormhole connecting from the errand thought, located a matter for consideration, a matter that is enticingly information-maximizing (broadly understood—we'll discuss this at great length in Part Three): Should we get new shoes?

We now examine the concept of getting new shoes and decide that it's the right thing to do.

Back through a wormhole to the errand plan, and we're ready to go.

We have reached a level of satisfaction that this errand plan optimally maximizes the order in our life. We will be less disordered. We will have lower entropy. We will have increased information content.

Introduction to P-Adic Arithmetic

P-adic numbers can be written looking just like decimal numbers: 873.14, for example.

If this is a p-adic number—with p = 11, for example—then the 7 within 873.14 is telling us how many 11s (not 10s) are in the number, the 8 is telling us how many 121s (not 100s), the 1 is telling us how many elevenths (not tenths), and the 4 is telling us how many 121sts (not how many 100ths). This is (so far) just the familiar base 11 arithmetic you learned about in junior high school.

But there is one important distinction between p-adic numbers and our familiar real numbers (whether written in base 11, base 10, or any other base): P-adic numbers have a very different notion of size. You can, in fact, consider p-adic mathematics as conceptualized backwards.

A real number's size is based, most importantly, on its first digit, the digit furthest to the left. For the real number 873.14, for example, the most important number indicating its size is the 8, which indicates that 873.14 falls within the 800s.

In p-adic mathematics, however, the furthest digit to the left is not what creates p-adic mathematics' analog to a real number's size. The furthest digit to the right does.

When we talk about a real number's size, we're probably referrring to its absolute value. For a positive real number, such as 873.14, its absolute value is 873.14. And -873.14, a negative real number, also has 873.14 as its absolute value. The size of either the real number 873.14 or the real number -873.14 is 873.14.

Mathematicians, who like to consider how other hypothetical

number systems besides real numbers might act, consider absolute value to be just real numbers' specific implementation of a more general concept, called the norm. And the norm of p-adic numbers is very different from the norm of real numbers.

For p-adic numbers, the norm—the analog to real numbers' absolute value—has to do with the p-adic number's last digit, 4 121sts for the 11-adic number 873.14. So the p-adic number 873.14 is not very close to 870, but is the same size as 319,498,207,695.08. Note that it's not even the value of the last digit on the right that's important in comparing the size of two p-adic numbers; it's just the fact that these two 11-adic numbers are both written with two digits to the right of the decimal point.

In the p-adic numbering system, we aren't always that interested in the digits far to the left, just as in our real decimal system, we aren't always that interested in the digits far to the right. In p-adic mathematics, two numbers based on the same prime number p are the same size (more technically, have the same norm) if the two p-adic numbers continue with the same number of digits to the right.

Perhaps more importantly, the p-adic numerals themselves are just names, not an indicator of size. As we've said, the p-adic number 4 two digits to the right of the decimal point is the same size as the p-adic number 8 two digits to the right of the decimal point. This 4 and this 8 are just two different names, two different labels, for objects of the same size.

The size of the 11-adic number 873.14 is 121. (We'll explain this later, but you might be able to guess that it's because two digits to the right of the decimal point give us a size or norm of 11^2 or 121.) This is different in a few ways from the size of the real number 873.14, which is 873.14.

For one thing, the size of a real number depends on all of its digits, whereas the size of the p-adic number depends only on the last digit to the right: The real number 873.14 is larger than the real number 162.54, but the two numbers are the same size p-adically.

For another thing, each digit of a real number represents a size: The real number 800 is bigger than the real number 700. In contrast, the p-adic number 800 is the same size as the p-adic number 700. In p-adic mathematics, two different numerals in the same position (same position to the left or the right of the decimal point) are just different names or labels for objects of the same p-adic size.

Finally, the digits to the far right don't make much difference for a real number's size: 873.14 is just a tiny bit bigger than 873.1. But p-adically, 873.14 is of size 121, whereas 873.1 is of size 11.

So real numbers give a lot of information about size. All the numbers matter. Real mathematics is metric and is focused on size.

P-adic mathematics is ultrametric and provides little information about size. The size of a p-adic number is given by the location of its last digit to the right, and it doesn't even matter what that digit is, just where it's located.

The size of a p-adic number is an indicator of its level of detail. The real number 873.14 is, for many purposes, only inconsequentially larger than the real number 873.1. But p-adically, the 11-adic number 873.14 is an order of magnitude larger (11 times larger) than the 11-adic number 873.1, because the p-adic norm (size) indicates the level of detail.

But what's the point of the rest of the digits of a p-adic number, aside from the location of its last digit? What, if anything, do p-adic numbers offer informationally to compensate for their very limited provision of information about size?

Here's the big surprise: P-adic numbers operate in a vast informational space. It will take the discussion in Part Two to explain how vast this is, how much information each p-adic number captures, and how p-adic numbers operate to provide an extraordinarily rich amount of information.

Right now, we can only hint at the beginning of this story, by examining what information the p-adic number 873.14 gives us, if all it tells us about its size is that it's size 121.

The 11-adic number 873.14 represents an object or entity that is called or labeled 4 and that is size 121. This object, 4, encloses or is preceded by an object or entity called or labeled 1, which is size 11. Object 1 (size 11) is enclosed or preceded by object 3 (size 1), which in turn is preceded by object 7 (size 1/11) and object 8 (size 1/121).

So p-adic numbers give us very limited information about size. After all, they confess to being ultrametric. Instead, p-adic numbers give us an ordered string of labeled events.

P-Adic Geometry

P-adic geometry, as you might imagine, also has its peculiarities. P-adic geometry involves only closed geometric shapes. And in p-adic geometry, either one shape is completely inside another shape, or else it's completely outside another shape, but two shapes never overlap in p-adic geometry.

In p-adic geometry, what matters is inclusiveness and the level of embedding or hierarchy. Shapes enclose whole levels of shapes, and shapes share their level with other shapes but never partially overlap with them.

P-adic mathematics' hierarchical geometric structure relates directly to the p-adic arithmetic that we've discussed. P-adic numbers that contain more digits to the right geometrically incorporate p-adic numbers with fewer digits to the right. A p-adic number with many digits to the right is large, and it is portrayed geometrically as incorporating (encircling) smaller p-adic numbers, which are numbers with fewer digits to the right.

P-Adic Mathematics: The Mathematics of Enclosure

P-adic mathematics is an astoundingly efficient way to record hierarchical structure. Said another way, p-adic mathematics is the mathematics of enclosure.

The number of digits in a p-adic number tells us how many levels of enclosure there are. The value of p for a particular p-adic number tells us what is the number of different labels or names that are possible at each level of enclosure.

The 7-adic number 5 represents the single enclosure called 5, which is one of 6 possible enclosures of the same size (1, 2, 3, 4, 5, or 6), or else there is the 7th possibility—0—which indicates no labeled enclosure of this size.

The 7-adic number 35 represents the object called 5 enclosing, or preceded by, a smaller object called 3.

The 7-adic number 635 represents a 3-tiered object: the larger object called 5, enclosing a smaller object called 3, which in turn encloses an even smaller object called 6.

The 7-adic number 4 represents an object that is the same size as 635, but is a different object—object 4—and has no enclosures.

The 7-adic number 205 represents a two-tiered object: the larger object called 5, enclosing only the very small object 2.

P-adic numbers map hierarchical structure, ordered structure, an ordered string of events or objects. Thinking p-adically means thinking about hierarchy, about what's enclosing what. Unlike real mathematics, p-adic mathematics is not measured. In math-speak, real mathematics is metric, but p-adic mathematics is ultrametric. P-adic mathematics notes only levels of hierarchy. In p-adic mathematics, two items at the same level of hierarchy are the same size as each other. P-adically, no measurement distinguishes items within the same level of hierarchy.

Not a Completely Novel Thought

The use of p-adic mathematics to model thinking is not unique to Pitkänen or TGD. A number of scientists studying the brain and the mind, even several physicists, have explored p-adic mathematics as a way of modeling thought processes. Several have even, like Pitkänen, explored the connections between real mathematics and p-adic mathematics.

What is completely novel of Pitkänen is incorporating p-adic mathematics and its relationship with real mathematics at the heart of a new theory of radical physics, a theory that brings consciousness and the mind to the heart of physics, and a theory that creates a physics of biology that shows living systems—biosystems—in constant dialogue with the real and conscious worlds.

Adele

P-adic mathematics can be thought of not just one p-adic number at a time, but rather for all possible values of p at once, every time you think p-adically.

Mathematicians use the term adelic mathematics to refer to the simultaneous use of both real mathematics, and p-adic mathematics for all values of p. An adele captures both a number's real representaion and the number's p-adic representation for all prime numbers p. There is a notation that is sometimes adopted to represent a number, as it is fully described on both its real basis and its p-adic basis:

$$(x_\infty; x_2, x_3, x_5, x_7, x_{11}, \ldots)$$

Here a number x is fully represented in its adelic form by its real representation x_∞ before the semicolon, followed by all aspects of its p-adic representation—its p-adic value for the prime number p=2, its p-adic value for the prime number p=3, and so on. (The notation x_∞ is sometimes used to represent x as a real number. It is intended to contrast the p-adic fields based on prime numbers p with the field of real numbers based on the "prime at infinity.")

TGD is at its heart adelic, in its simultaneous understanding of every object and every phenomenon through both its metricized real representation and its p-adic representation of enclosures.

According to TGD Digital Mind Math, the mind develops an intricate labeling procedure based on p-adic mathematics. This labeling procedure maps an intricate interconnectedness among our thoughts. The mind's tendency is to maximize the p-adic size of our next thought, either by increasing the complexity of the previous thought (higher p) or by migrating to a thought with more levels of complexity (higher negative-n; more digits to the right).

The mind is very smart, very forward-looking, capable of considering possibilities many thoughts into the future. As a result, the mind's maximization of information content may be determined not just by the immediate next thought, but perhaps based on a master plan, many thoughts into the future.

This is similar to the concept in financial analysis of present value of future profits. Perhaps a profitable project in the long term will require a short-term capital expenditure that will depress the immediate profitability. This capital project will still be profitable if the present value of future net cash flows exceeds the short-term capital expenditure.

In the same way, the mind, using p-adic mathematics, and drawing from our individual life experience to date, continually models a large configuration space of possible next thoughts, then selects an information-maximizing thought-line to proceed along, which we do, as our physical being momentarily collapses to the real experience whose p-adic label the mind has provided.

4.   Our Blinking Moments of Time

Taking Time Seriously as a Fourth Dimension

Time has been pretty well ingrained as the fourth dimension. In fact, this is so well ingrained that the Twilight Zone and other science fiction need to start us off at the fifth dimension if they want to take us somewhere really weird.

Up until now in this book, we've been a bit glib about assuming time as a fourth dimension, discussing TGD's four-dimensional spacetime sheets, and adding four microscopic dimensions to create TGD's eight-dimensional embedding space.

But what do we really mean when we contemplate time as a fourth dimension?

The Macro-Extended Dimension of Time

When we look at an object, its three spatial dimensions are apparent. In a traditional (not TGD) understanding of space, the object extends some way along the universe's macro-extended north/south dimension, some way along the universe's macro- extended east/west dimension, and some way along the universe's macro-extended up/down dimension.

TGD has a different way of looking at the object's three dimensions. In TGD, the object is its own universe, and its extension In three spatial dimensions creates the spatial boundaries of its universe.

But neither this traditional description of the three spatial dimensions, nor this TGD description of the three spatial dimensions, has yet introduced time as its fourth dimension.

In the traditional understanding, time extends macrotemporally, just as the three spatial dimensions extend macrospatially: Time extends from the Big Bang to the end of the universe. So a four-dimensional understanding of an object is the spatial extension of the object, as it varies over time from the Big Bang to the end of the universe.

Similarly, a TGD understanding of a four-dimensional object is the object's spatial extension, from the beginning to the end of its existence, as it has changed from its inception, through all stages of its existence, until its demise. This is TGD's four-dimensional spacetime sheet, the whole spatial history of an object from its inception to its demise.

A TGD object—a TGD spacetime sheet—is a universe bounded spatially and bounded temporally. The temporal bound of this object's spacetime sheet is the length of its existence, from its creation in the past to its demise in the future. And the spatial bound is its changing spatial extension, as it has varied in the past and will vary in the future. A TGD spacetime sheet is this four-dimensional history and future of an object, a four-dimensional universe, with its varying spatial bounds during its temporally bounded existence.

The Micro-Extended Dimensions

TGD spacetime has eight dimensions, four macro-extended, and four micro-extended with just enough extension for tiny wormholes.

The micro-extended dimensions of TGD are extended spatially only (no time dimension; just four spatial dimensions). The spatial dimensions are very small—approximately 10^{-30} meters—and are therefore not perceptible to our human senses. In the Topological Geometrodynamics model of the universe, they do, however, really exist, and are an integral part of TGD physics.

Wormholes reside within the micro-extended spatial dimensions, connecting macro-extended spacetime sheets. As time advances macrotemporally, wormholes remain in place, or are newly introduced, or are extinguished. In other words, not only do macro-extended spacetime sheets change spatially at different moments of time, they also change their enclosed and enclosing wormhole connections.

So TGD's eight dimensions record spacetime sheets as they change spatially or remain constant from moment to moment, and as their hierarchical wormhole links change or remain constant from moment to moment. P-adic mathematics keeps track of the evolving reality.

A Moment of Your Time

TGD analysis suggests a tenth of a second as a fundamental biorhythmic time scale. Dr. Pitkänen discusses this in the Afterword. During each brief TGD moment, this blink of an eye, a lot happens. This blink incorporates TGD's cognitive quantum jump, during which an intricate set of cognitive possibilities is generated, then—guided by the negentropy (information) maximization principle—is reduced to a single experienced path.

In TGD and Digital Mind Math, this quantum jump—the core process of thinking, the mind's core procedure—consists of three steps, starting at the last experienced thought:

- The mind <u>generates</u> an intricate set of <u>possibilities</u> for its next thought. This set of possibilities is a configuration space, a world of classical worlds, a set of possible candidates for the one real thought or event that we will next experience. Each of these possibilities—labeled p-adically, taking advantage of p-adic mathematics' great organizing capabilities—is anchored at the most recently experienced moment. Using terms from physics, this quantum world of all that is possible, this sum over histories, is a world of classical worlds, an enormous set of imagined but not experienced possibilities for the next thought, for the next real experience. But it's only a configuration space, only the set of hypothetical next experiences, not an experience itself.

- Next, guided by the principle of maximizing information content, the mind <u>selects</u> which single path, from among all the possible paths generated, will be the path that we will actually experience. This selection process takes advantage of the p-adic norm, p-adic mathematics' definition of size: To maximize information content, the mind selects the path that is labeled with the largest p-adic number (high number of details; many levels of complexity). Here when we refer to the largest p-adic number, we mean, of course, the number with the largest p-adic norm, the largest p-adic size, according to how p-adic mathematics defines the norm and the size.

- Finally, we <u>experience</u> the selected path. We experience the thought that has been selected. This is a momentary real experience of the thought that until now has only been analyzed p-adically. We have reduced the configuration space of possible thoughts to a single path, a single thought, which we now experience. After this momentary experience of the selected event, we are immediately ready for our next blink, our next moment of generation of a new set of possibilities which, after selection of the information-maximizing path, we will once again reduce to a new single path that will again form a momentary experience.

TGD looks closely at this moment, this blink, this quantum jump. In TGD, the quantum jump is not a single indivisible process. In TGD, the quantum jump is a series of rapid processes, always following the same pattern: generate possibilitie; select; experience.

Quantum physics' reduction to a single real path results in an unstable condition that must immediately be brought once again to the world of quantum possibility and the generation of all possible realities that are consistent with the previously selected and experienced single path.

Our experience of time is our sequence of quantum jumps. This is who we are.

We continually experience selected events, based on the principle that the selection is guided by the maximization of information content. From all possible next paths, the single next path that is selected is the path with the maximum amount of information.

Information content sounds like it belongs in a field outside of physics, but surprisingly it is a well-established quantity of mainstream physics.

In mainstream physics, entropy is a central concept, which represents a system's degree of disorder. Disorder—entropy—is a well-defined concept of mainstream physics, a measure of randomness, of absence of pattern, of absence of information.

Pitkänen, along with some other physicists, emphasizes the negative of disorder—the negative of entropy—which is called negentropy and which is a measure of information. In TGD, the quantum jump's reduction to a single path is guided by the Negentropy Maximizaton Principle (NMP), the principle of mazimization of information.

$M^4_+ \times CP_2$

One key to TGD's model is a reconfiguration of the physical model of spacetime. This reconfiguration then becomes the foundation for TGD's applications to physics, biophysics, and the mind. The reconfigured spacetime is one that permits easy application of p-adic mathematics, because TGD's spacetime is hierarchically orgainized as enclosings of standard physics' four-dimensional Minkowski spacetime M^4_+.

Each TGD object is a bounded four-dimensional universe, extending in three spatial dimensions, and extending in time from its creation to its demise. Minkowski spacetime M^4_+ is contemporary physics' entirely standard way of referring to these four dimensions of space and time:

- It is four-dimensional, incorporating time, in addition to three dimensions of space.
- It is Minkowski space, which means that space's three dimensions are not precisely perpendicular to each other, because the axes must be adjusted for special relativity. Special relativity is the earlier aspect of Albert Einstein's theory of relativity. Special relativity is an entirely standard aspect of all modern physics, uniformly accepted, and the mandatory model for space and time. According to special relativity, space has three precisely perpendicular axes only for objects that are not moving with respect to each other, objects that have zero relative velocity. To allow for any relative motion at all, we must permit space's three axes to be adjusted based on the object's velocity. Physicists routinely use Minkowski space as the model of space; it reduces to ordinary space, with three mutually perpendicular spatial axes, if and only if an object is not in motion. An analogous analysis brings time into the model, also in a way that varies according to the object's velocity.
- The plus sign means that both space and time are measured from the Big Bang.

In TGD, spacetime is not simply four-dimensional M^4_+ spacetime, with objects situated at particular coordinates within an M^4_+ spacetime that surrounds the object. Instead, each TGD object, starting at the elementary particles, is its own macroscopic and macrotemporal event in bounded M^4_+ spacetime. An atom also exists as its own bounded M^4_+ spacetime, linked to the elementary particles that it incorporates by a microscopic four-dimensional wormhole.

Operating in four-dimensional space gives these wormholes enormous power to connect. These CP_2 wormholes don't determine just one and only one connection from an M^4_+ four-dimensional spacetime sheet to a second M^4_+ spacetime sheet. The four dimensions of CP_2 space permit wormholes to connect an M^4_+ spacetime sheet to many other M^4_+ spacetime sheets, to an unlimited number of M^4_+ spacetime sheets, to whatever other M^4_+ spacetime sheets that it encloses or that enclose it.

A given M^4_+ spacetime sheet can have many conceptual frameworks, in each of which it is enclosed in a different series of larger spacetime sheets, and in each of which it encloses a different series of smaller spacetime sheets. The wormholes' four-dimensional space creates this multiplicity of enclosing relationships, and this multiplicity of enclosed relationships.

The $M^4_+ \times CP_2$ world is an intricately interconnected world of worlds.

So a smartphone, for example, is a macrospatial and macrotemporal M^4_+ object, extending along its three dimensions of space and extending temporally along the fourth dimension, time. A smartphone is a type of computer. In Digital Mind Math, this means that the M^4_+ spacetime sheet that is the concept of a computer encloses the M^4_+ spacetime sheet that is the concept of a smartphone. The smartphone spacetime sheet condenses to the larger computer spacetime sheet by a wormhole contact identifying the relationship.

When we are generating possibilities related to the thought of a computer, one set of possibilities will involve various types of computers, branching through any wormhole that identifies concepts, such as smartphones, that computers enclose. This is how Digital

Mind Math operates, taking advantage of its efficient p-adic structuring, which we will learn much more about in upcoming chapters.

But a smartphone is not just a type of computer. It's also a type of telephone.And a type of email device. And something you put in your pocket. Maybe a status symbol. And so on.

So the smartphone concept spacetime sheet is not limited to being a type of computer. A smartphone does not appear only in the one conceptual hierarchy, in the computer category.

Four-dimensional CP_2 space permits unlimited hierarchies for the smartphone concept to be part of. The smartphone is a p-adic digit to the right of the p-adic computer digit—within the computer category, a specific example of a computer.

But the smartphone concept is also part of another p-adic string, the telephone string. The same conceptual object—the smartphone—is part of many different conceptual hierarchies. Mathematicaly, Digital Mind Math permits this, because the wormhole connections CP_2 are four-dimensional.

We will continue in upcoming chapters to detail the Digital Mind Math processes and to give examples of Digital Mind Math in action, as the mind uses it, as we think.

TGD's eight-dimensional space is not just any eight-dimensional space. It is the specific space $M^4_+ \times CP_2$. This is the mathematical space of four-dimensional objects, which are connected by four-dimensional wormholes.

TGD's hierarchical structure for the real physical world is straightforwardly mapped p-adically, since p-adic mathematics is the mathematics of enclosure, and since TGD's $M^4_+ x CP_2$ space is all about levels of enclosings and enclosures.

Spacetime 4-Spheres, Intricately Linked

As a synonym for the mathematical expression $M^4_+ \times CP_2$, we'll be using the expression spacetime 4-spheres, intricately linked. Let's see why.

Spacetime, a term from modern physics, means the combination of the three dimensions of space (up/down; left/right; forward/backward) plus the fourth dimension, time. Modern physics has been forced to work with the single concept of spacetime, rather than the separate concepts of space and time, ever since Albert Einstein developed special relativity—the early part of his theory of relativity—early in the twentieth century.

The later part of Einstein's theory of relativity, called general relativity, is a theory of gravity, which even today, almost a century after its publication, has not been fully reconciled with modern physics' other main strand, quantum physics. There is still no single uniformly accepted unified theory that unites general relativity with quantum physics.

But special relativity—the theory that moving clocks run slow, and that objects contract along the direction that they are traveling—is a proven theory, precisely confirmed by experimental observation. Special relativity is a standard part of modern physics, and no modern theory can be put forth without automatically assuming special relativity.

A Minkowski diagram, developed by the German mathematician Hermann Minkowski, is a popular way to display the effect of special relativity.

When an object—let's call it object #1—is moving, a clock traveling along with object #1 (said to be in object #1's "proper frame") will tell time normally, and a ruler traveling in object #1's proper frame will measure distance normally. But—and this is the surprise conclusion of special relativity, now uniformly accepted—a clock standing still will not measure time the same as the clock that is traveling with object #1, and a ruler standing still will not measure distance the same as the ruler that is traveling with object #1.

Even though things look normal within object #1—let's say it's a spacecraft traveling on a long mission—time and distance are not the same compared to objects back on Planet Earth. This explains the Twin Paradox: The twin who stays on Earth is older than her astronaut twin sister when the astronaut returns home.

Hard to believe or not, special relativity is a fully verified and accepted phenomenon, and physicists routinely use Minkowski spacetime and Minkowski diagrams—which adjust the relative angles of the three dimensions of space, depending on the relative speeds of the two objects, so that the three spatial dimensions are not precisely at 90-degree right angles to each other. Minkowski spacetime makes a similar adjustment for time.

So M^4_+ is a spacetime 4-sphere, in the sense that it correctly shows three-dimensional objects as they vary in size and shape along a fourth dimension, time—correctly, in the sense that it has made the needed adjustments for special relativity.

The M as well as the + sign are just technical details: The M means we're smart enough to adjust for special relativity. And, to keep things simple, both time and space are measured starting from the Big Bang, so that this way all times and distances are positive numbers.

But the essence of the spacetime 4-sphere M^4_+ is that it is a three-dimensional motion picture: objects with three spatial dimensions, passing through time.

A little jargony perhaps, but this is physics and mathematics, plus we're not done yet: We still have to discuss dimensions five, six, seven, and eight. But the first four dimensions—M^4_+, the spacetime 4-sphere—give us the powerful concept of a shorthand way to think of a memory.

When we watch Johnny cross the street, we form an M^4_+ spacetime 4-sphere, and store it in our brain. Later on, we can retrieve this M^4_+ spacetime 4-sphere and replay it.

But replaying this M^4_+ spacetime 4-sphere—Johnny crossing the street—is not the only thing we can do with this memory. When we store this memory, we automatically link it with concepts and memories that are already there—thoughts about Johnny, thoughts about the street scene, thoughts about the time of day.

And when, later on, we read that the First National Bank was robbed, and we create and store the fact of this robbery as an M^4_+ spacetime sphere memory, we realize that we have a memory for that street at that time: It's exactly where and when we saw Johnny cross the street!

We automatically link new memories and thoughts in multiple directions to all sorts of existing memories and thoughts, in ways that allow us to link anything to anything that it relates to, intricately linking whole thoughts and parts of thoughts to other thoughts and other parts of thoughts—intricate interconnections.

Digital Mind Math's and Topological Geometrodynamics' mathematical model for these links is as tiny four-dimensional wormholes, existing in four-dimensional CP_2 space, linking the macrospatial and macrotemporal M^4_+ spacetime 4-spheres that are our thoughts and our concepts and our memories.

Because the four-dimensional CP_2 wormholes operate in a separate four-dimensional space from the macroscopic and macrotemporal M^4_+ spacetime 4-sphere concepts, the wormholes are able to infinitely link any concept with any other concept: spacetime 4-spheres, intricately linked.

This is the structure of Topological Geometrodynamics' physics as well as the Digital Mind Math model of the mind—$M^4_+ \times CP_2$ spacetime 4-spheres, intricately linked. Thoughts and concepts and memories, interconnected with every other thought and concept and memory that it relates to.

And the punchline of all of this is that p-adic mathematics— the natural mathematics of cognition—is the natural mathematics to imagine and keep track of and analyze and label and retrieve all of these thoughts and concepts and memories. P-adic mathematics is the mathematics of enclosure, and as such is the numbering system for the spacetime 4-spheres of memories and thoughts, along with all of their intricate interconnections.

5.   Review of the Basics

We have now completed the foundation of Digital Mind Math, the basic organizing principles of the mind, according to the theory of modern physics TGD, Topological Geometrodynamics.

Let's review.

In your day-to-day existence, your physical body proceeds through space and time, in four-dimensional macrospcopic, macrotemporal spacetime.

This day-to-day processing can be understood through the laws of physics, traditionally pictured as your four-dimensional body occupying changing segments of infinitely extended three-dimensional space, while the fourth dimension—time—marches on.

In TGD—in contrast to traditionally accepted physics—your body is a universe, finitely extending to its outer boundary.

Your body is its own universe, enclosing other universes, enclosed by still other universes—four-dimensional universes enclosing four-dimensional universes throughout eight-dimensional embedding space.

These four-dimensional universes are microscopically connected in CP_2 space.

CP space means complex projective space. We won't be going into the mathematical details, except to note that CP_1 space has two real dimensions, and CP_2 space has four real dimensions.

CP_2 space has four dimensions. And it is its fourth dimension that gives CP_2 space all of its power for multiple connections in many directions.

Just to be clear, we had earlier established that just one extra dimension—a fifth dimension—would be enough to allow an infinite hierarchy of M^4_+ spacetime sheets. But one extra dimension would permit just one single connecting line from one sheet to just one other sheet—just a line, without any left/right or front/back thickness, and just a single line of connections from one sheet to another sheet to another.

So a fifth dimension, while permitting an infinite hierarchy of spacetime sheets, wouldn't permit a very interesting infinite hierarchy: (1) there would be no physical reality to the wormhole, since one dimension allows just a line, with no extension to the left and right, and no extension to the front and back; and (2) there would just be a single line of connections possible, along just the one dimension.

The sixth and seventh dimensions of $M^4_+ \times CP_2$ space give each wormhole a physical reality—a left/right dimension and a front/back dimension, in addition to the connecting length that the fifth dimension offers. And the eighth dimension gives the wormholes the capability to infinitely interconnect, so that they do not operate solely along a single line of wormholes, but instead can connect along multiple paths, infinite paths—that is, to intricately interconnect.

TGD physics takes place in $M^4_+ \times CP_2$ space, the four-dimensional Minkowski spacetime M^4_+ of modern physics (positive time and positive space only), crossed with a small four-dimensional connecting space.

In real TGD physics, real four-dimensional finite universes contain and are contained by other four-dimensional universes. Real microscopic four-dimensional wormholes connect the real macroscopic and macrotemporal physical objects, implementing the structure of encloseds and enclosings.

There is a second way to make sense of TGD physics—a second way, besides as real physics. This second way is to understand TGD physics p-adically.

P-adic mathematics tracks physics only organizationally, not metrically.

The p-adic understanding of TGD physics tracks the enclosings realized by TGD physics' wormholes.

P-adic physics is the physics of the mind—of thinking, sensing, analyzing, dreaming, regretting, re-creating.

Digital Mind Math, the p-adic-based mathematics of the mind, permits us to think back toward the beginning about what was and what might have been, and think forward toward the future about what will be, won't be, can be, can't be.

Our Blinking Existence

Our blinking existence moves us along temporally—in small temporal jumps, in the direction that runs from the Big Bang toward the end of the universe—with each jump maximizing the information content of our existence.

At each moment of time, our mind, taking advantage of p-adic mathematics' great informational space, generates a multitude of possibilities, each labeled p-adically, each systematically, even simply, generated as obvious possibilities for what to think or do next.

P-adic mathematics is a natural framework that the mind is structured to use based on the evolutionary advantage provided by p-adic mathematics, the great efficiency of p-adic mathematics.

We do not have to know about p-adic mathematics in order for our minds to use p-adic mathematics. We speak prose even if we do not know what prose is.[2]

Not only is p-adic mathematics the natural mathematics of mapping hierarchies and organizational relationships, p-adic mathematics also gives us a straightforward and natural measurement of information. The larger a p-adic number is (using the p-adic definition of size), the larger the amount of information.

So Digital Mind Math uses p-adic mathematics to generate possibilities, and also to make the choice as to which possibility best maximizes information content.

Then we blink.

Then we experience the selected thought, the thought, as p-adically labeled, from among all the p-adically labeled possible thoughts that we generated p-adically. Our selection is based on our p-adic calculation of information maximization.

Finally and momentarily, after this p-adic analysis (p-adic idea generation and p-adic idea selection), it is a real thought that we experience. We open the file drawer, labeled p-adically, that contains the selected information-maximizing thought, and find the real recipe for this thought, the recipe that allows us to experience this thought, the instructions that trigger a set of real visual and auditory and other sensory experiences, that trigger body movements and other sensations, that put us on a holodeck of time-ordered real experience.

In that moment of time, we realize our next thought.

P-Adic Efficiency

Even if you're convinced of the power of TGD's $M^4_+ \times CP_2$ space of encloseds and enclosings, it may seem hopelessly complicated as a mathematical tool, until you're also convinced of the great efficiency of p-adic mathematics.

Remember that p-adic mathematics is not about capturing a single set of enclosures. It is about capturing at once all sets of enclosures—the entire universe of universes of each object or concept or thought, as it interrelates to each of the other objects or concepts or thoughts in its universe of enclosures.

We will describe this great p-adic efficiency in Part Two.

But first, a little more practice with the Digital Mind Math framework.

6.   Memories

What is a memory? Where do memories reside? What triggers a memory?

Let's briefly digress to see how Digital Mind Math answers these questions.

In Digital Mind Math, connections among the neurons in our brain have developed over the course of our lifetime.

Each previous thought, and each previous action, has made its mark, left its trail, in the connections among our neurons.

According to our evolutionarily developed process of thinking, right after our most recent thought, the neural system examines the now-revised neural relationships and tends toward a next thought that decreases entropy, that increases information content.

This is a natural function of the brain, which, upon examination, takes place by the brain's natural use of p-adic mathematics to find a p-adically large next path.

The p-adic map of the brain's neural structure is incredibly elaborate and far-reaching. A map of the brain's neural structure must allow for the possibility that any neuron can connect to any other neuron. Digital Mind Math constructs this map, which permits rapid identification of information-maximizing routes.

This incredibly elaborate and far-reaching task is easy for p-adic mathematics to accomplish, which is why our brains have naturally evolved to operate p-adically.

A memory is represented in the brain by a specific ordered neural sequence.

Each element of the memory is multiply connected to other ordered neural sequences.

Information content increases by making connections more detailed, or by making new connections.

When the brain biologically selects for the mind what will be the mind's next thought, this thought—labeled as the selected sequence of neurons—is realized, is experienced by the person, through his or her brain.

The experience, labeled p-adically by the organization of the brain's neurons, is experienced through the real corporeal sensations that correspond to the selected path.

We have found this memory because it solves a problem for us, or because it enhances our personhood, or because it helps to clarify a course of action.

Digital Mind Math assumes a hierarchically configured organization of time and space, which p-adic mathematics is very good at recording and labeling.

The elements of the selected neural pathway are each linked multiply to other pathways, the elements of which are also multiply linked.

For the path selected to experience, visual sensations and aural sensations are activated, even though there is not an external physical source for these images or sounds.

The selected neural pathway—the thought, the memory—is also linked to endocrine functionality that triggers emotional activity.

And the selected neural pathway is linked to motor control functionality that triggers movement, or action, or the sensation of movement or action.

A Guess As to How the Brain Rapidly Identifies Information-Maximizing Neural Routes

As previewed in the Introduction, at times we will proceed along an Occam's Razor route to suggest how the Digital Mind Math mind could fit right onto the brain.

In other words, here is a simple physical, biological, neurological process that the brain may follow in order to identify information-maximizing thoughts:

Perhaps, the brain simply lets out a local charge, measures entropy-minimizing effects, and activates a low-entropy route.

This would generate what we would experience as an information-maximizing thought.

Too early

This neurological possibility is just speculation.

But even the Digital Mind Math model of a memory presented so far still only frames the picture of how Digital Mind Math works. We have yet to present the complete repertoire of mechanisms that give credence to the claim that Digital Mind Math is the model of how we think.

This will be established piece by piece, by example and by theory, starting with another sample application in the next chapter.

7.   The P-Adic Surgery of Dreams

Digital Mind Math describes dreams precisely.

The process leading up to a dream is the core process of quantum physics—the quantum jump—applied in the setting of nocturnal cognition.

Leading up to a dream, our mind, loosened by its descent into sleep, goes through Digital Mind Math's three-part cognitive process:

First, the mind generates possible next thoughts, possible dreams. These possibilities begin at important events, pivot from important events (real events, or psychological events, even imagined events), and reach backward and forward in many directions, as candidates for our dream.

As is the case for our waking thoughts, each possible dream is intricately interconnected to emotional categorizations, labeling of unresolved issues or personal goals, work-related projects, process improvements that will make our life better, inventions and discoveries that will improve our local world or a larger community.

This dream-generation process takes particular advantage of a loosening of the mind from its literal grip on reality, which the sleep world permits us.

As a result, the possible dreams that are generated are able to take advantage of what could be labeled p-adic surgery—a thought segment, which has a basis in a memory, but which has been surgically revised to change a specific action or comment or face or person or location.

How easy this is for a TGD brain to do. The TGD brain has access to millions and millions of memory strings, which are records of time-ordered events, consisting of ourselves as enclosing universes enclosed in interaction with other universes.

With dream possibilities generated—many with fanciful substitutions of richly imagined aspects for reality-based narratives—

our mind moves on to the second of the three-part Digital Mind Math cognitive processes: selection of the dream, guided by information maximization.

Here our mind measures, through a strictly mathematical (p-adic) process, the information content of the candidates for dreams.

This chapter is just an introduction to the process, an overview to set the stage and get the reader's mind running in the right direction. The details—the mathematics of possibility generation and information-maximizing selection—will unfold through examples throughout Part Three of this book, after we ground ourselves more firmly in p-adic mathematics in Part Two.

But for now, imagine that our minds are able to rapidly develop a host of possible next thoughts—possible dreams—mechanically pivoting off of existing thoughts, influenced by our history of prior thoughts as they have shaped the map of our interconnected neurons, taking special advantage of the sleep state's loosening of the grip on reality to allow implausible surgical substitutions.

How will each possible dream resolve bothersome emotional problems? Make our life happier? Envision achievement of our goals? Solve a practical problem?

Digital Mind Math makes this easy.

Not only, at Step One—idea generation—does Digital Mind Math reduce this to a straightforward mathematical process. At Step Two also—idea selection—Digital Mind Math makes this mathematically straightforward, as measured by the p-adic norm, the informational size of the idea, the magnitude of the dream's information content.

Again, this is just stage-setting, outlining, of the Digital Mind Math framework. Just general discussion of how Part Two's deeper understanding of p-adic mathematics will be applied to numerous real-life examples in Part Three.

So imagine that the mind has generated many dream possibilities, and selected an information-maximizing approach. Now, go ahead. Dream. Experience the selected dream.

We now move away from p-adically labeled dream possibilities, and p-adically analyzed dream selection, to the actual experience of the dream.

The selected dream takes its p-adic label from the configuration space of possible dreams, and is now experienced in its full real actualization.

The selected dream path connects to paths with visual and aural instructions, other sensory associations, factual contexts, emotional connections.

We experience the dream, which has been selected as information-maximizing from among the possibilities that we generated, possibilities based on the history of prior quantum jumps which have formed our brain's neural network, possibilities creatively and surgically modified to explore the implausible, the novel, the hypothetical.

Now wake up and think about that dream. It's time for a next thought.

Can the mind pivot off last night's dream for a next round of information maximization?

Daydreaming Too

Of course, the process of daydreaming—the creative switch away from attending to hard reality—looks just like our night dreams.

Daydreaming satisfies an urge to redirect attention from the instant reality, the current problem, and imagine other possibilities.

Daydreaming deploys p-adic surgery to switch aspects of the familiar, measure the possibilities, and experience a vision of an alternate course.

If it's successful, the next thought after the daydream will pivot off of this high-information alternative.

Connecting to the Past; Connecting to the Future

Whether in a nighttime dream or a daydream, our mind is particularly adept at generating thoughts not only based on the past, but also projected into the future.

M^4_+ spacetime thought bubbles extend as temporal segments in addition to spatial segments. An M^4_+ spacetime 4-sphere proceeds from its temporal beginning to its temporal end.

As we generate possible next thoughts, these possibilities can take many temporal forms: We can relive a past moment realistically. We can relive a past moment, surgically reconstructing it. We can use a past event as a springboard into possible futures. We can construct purely hypothetical futures.

Just as, in a possible next thought, we may spatially reduce or spatially expand—compact ourselves into a litle ball, or jump higher or stretch further than we ever have—so too may the temporal span of of our possible next thought have temporal boundaries entirely in the past, entirely in the future, or crossing from past to future.

A night dream or a daydream emerges as the thought selected to experience because it is evaluated as maximizing information content. The selected thought might be real or hypothetical, and might be in the past, present, or future.

8.   Seconds, Minutes, Hours, Days, Weeks, Months, Years

We've discussed p-adic arithmetic and p-adic geometry, but we haven't yet shown how this strange form of mathematics relates to our more familiar real mathematics. This correspondence between real mathematics and p-adic mathematics is a critical element of TGD, since it allows the mind to connect with the real physical world, formng the basis for TGD's mind/biosystem/physical world links.

In this chapter, we'll examine several ways in which we measure time, and consider how we sometimes think about these different measurements independently, and at other times are led to think about how these different measurements relate to each other.

The pattern of sometimes thinking separately about these processes—human-invented time, sun time, moon time—and sometimes thinking about how these processes relate to each other, is reminiscent of sometimes operating p-adically, sometimes operating with real mathematics, and sometimes taking into account the commonalities.

Some processes—generating possible next thoughts, and selecting the next thought based on information maximization—are best done p-adically. Other processes—the experience of the selected next thought—have real mathematical models. And the real mathematical models are recipes in a file cabinet labeled p-adically. There is enough information in common—Pitkänen proposes that the common information is represented by the rational number that is the common basis for both the p-adic number and the real number—to create a link from p-adic to real and from real to p-adic. (In the Afterword, Pitkänen elaborates on this, describing a strong form of holography, at the level of the world of classical worlds.)

Similarly, a common basis links the sun and moon calendars with human-defined calendars. For some purposes, we use human-defined calendars. For other purposes, we use sun- or moon-defined calendars. And at times we need to reconcile using algorithms deriving from a shared common basis.

The Real Calendars

Think back in time.

No. Not just to what you ate for lunch yesterday.

Further back. To before there were street lights and electric utilities. No computers or smartphones either. Just living in caves, spending a lot of time—day and night—outdoors.

Back in the cave era, our concept of days was solely based on the sun. A day begins at sunrise. It gets brighter. The sun sets and it's dark. Then another day begins at another sunrise.

This is a very real world, in which the real sun defines for us the boundaries of the day.

This day is based on the real world, not based on human-created concepts. Birds know about this day. And bees. And cave-dwelling humans.

The more clever cave humans also noticed that there is not just a daily pattern, but also a monthly pattern.

The shape of the moon, and the time of night it is first seen, and in what part of the sky it is first seen—these all change night by night. But the pattern repeats itself. Every month (sounds better than every "moonth").

And the really wise among the cave people caught on to an annual cycle too. The whole pattern of warmth and cold, of long days and short days, of sun high and low in the sky, is also cyclical, creating the year.

Our ancient selves had reality-based days, reality-based months, and reality-based years.

Over the centuries, these real, astronomy-based days were explained by the rotation of the Earth on its axis. And the real, astronomy-based months were explained by the revolution of the moon around the Earth. And the real, astronomy-based years were explained by the revolution of planet Earth around the sun.

Some argue that our ancient selves—grounded in the heavenly cycles of Mother Gaia—were more in touch with our place in the universe than we modern humans are.

But practicalities have intervened: Digital watches and clocks. Published time schedules for planes, trains, and television programs. Alarm clocks.

Yet we want to preserve some relationship with the astronomy-based concepts of days, months, and years.

As a result, we have learned to live with with a dual set of concepts—purely humanly-invented timeframes such as seconds, minutes, and hours, alongside the hybrid days, months, and years, which do have rigorously humanly-defined measurements, but which also bear a connection with real concepts based on astronomy.

The P-Adic Calendar

Sixty seconds equal one minute. This is a precise equivalence because this equivalence is purely a human concept, with no basis in physical reality.

Sixty seconds equal one minute because we say they do. This is a human invention, a concept, an idea, not a reality of cosmology or a reality of particle physics.

And sixty minutes equaling one hour is equally precise, and equally conceptual, again with no basis in physical reality.

An hour consists of exactly 3600 seconds, and each grouping of sixty seconds is called a minute. With no appeal to any physical reality, we may be very precise and we do not have to reconcile these concepts with any physical reality.

Trouble looms, however, when we jump to the twenty-four-hour day.

Suddenly, we have a physical reality. Successive sunrises—and successive sunsets, and successive high points of the sun—remind us that there's a physical reality to the concept of a day.

We desire a concept of day, because we don't want to count hour after hour after hour, without taking into account that, after about twenty-four of them, we start repeating astronomical realities.

We've lost our precision, though, because—no matter how many human-constructed segments we count from sunrise to sunrise—we've got a mess on our hands. Each sunrise-to-sunrise day has a different length than the previous or the next.

And worse, we don't want to count day after day after day, without taking into account that after 28 or so of them, we start repeating another set of astronomical realities, related to the moon. And after about 365 of them, we repeat astronomical realities related to the sun and the other stars.

Way too many moving parts here, so we begin compromising, begin introducing some manageable conventions for how time will be aggregated.

A day will be a fixed length, a human concept that is only irregularly based in physical reality.

The concept of day represents a precise measurement of 86,400 seconds. We'll even toss in a precise measurement of 604,800 seconds—seven days—and call it a week, another entirely human concept with no physical equivalence.

But we cannot—everywhere and eternally—make a year precisely 365 days.

A year is kind of a mess, actually.

If we want to keep seasonal patterns consistent—snow in the northern winter, hot days with the sun high in the northern summer—we need our human-defined conceptual year to bear a fairly uniform relationship to the real, astronomy-based year. Which means that one January 1st needs to be be just over 365.24 days after the previous

January 1st.

But this creates another mess.

We can't have January 1, 2016, begin at midnight, and January 1, 2017, begin just past 365.24 days later, at 5:49 in the morning.

So to keep New Year celebrations at midnight, and also keep the northern winters snowy, we introduce an extra day—a leap year— 97 times every 400 years.

Our desire for a mathematically straightforward concept of day and year conflicts with our desire for an astronomically reasonable day and year.

The humanly developed concept is difficult to reconcile with the physically real concept.

Months. Months are also problematic. In theWestern calendar, they're a hodgepodge of 28 or 29 or 30 or 31 days, with not much relationship to their etymologically related heavenly body, the moon.

The Western calendar basically punts on trying to keep the months moonly.

But some calendars—for example the Hebrew calendar—are lunar: The new moon occurs on the first day of the month.

No easy ride here either, though: First of all, the precise time of the new moon can fall any time of the day, so there is no uniformity about, say, the beginning of the first day of the month falling precisely at the new moon. The new moon is at some time during the first day of the month, but not at a predictably regular time of day from month to month.

The bigger problem that lunar calendars must address is how to keep some semblance from year to year in what the weather is like each month, how to keep a specific month generally in a specific season, not jumping all over the place, winter one year, summer another.

So the Hebrew calendar is on a 19-year cycle, with 12 years out of the 19 having 12 months, and 7 years out of the 19 having 13 months. This is how a lunar calendar can also match a solar calendar's ability to be seasonally consistent from year to year.

Many other calendars are in use today throughout the world, some based on the sun, some based on the moon, some based on astronomical observation, some based on astronomical data, some untying the link from human-labeled month to real climatic season. But no matter what the calendar's design and specifications, its users are forced to at times think with arbitrary human-defined concepts, and at other times think with real geo-astronomical concepts, and at still other times consider how these sets of concepts interrelate.

The Adelic Mathematics of the Calendar

When we try mixing the purely conceptual components of time—the components deriving from human consciousness, from the mind—with real, physical, astronomical concepts of time, we are using adelic mathematics.

Adelic mathematics is the mathematics of real and p-adic mathematics, the mathematics that understands the concept of number to have both real and p-adic components, the mathematics that meets the challenge that a complete mathematical understanding requires a number to be expressed by its full real expansion and its full p-adic expansion for all prime numbers p.

9.   Cognition, as Modeled by Piaget and Vygotsky

We have some p-adic mathematics ahead of us in Part Two, and many examples ahead of us in Part Three to put some meat on the bones of Digital Mind Math's p-adic structure. But perhaps this aside—to psychology, cognitive development, pedagogy, and epistemology—can serve as another warm-up exercise for the tasks ahead.

Cognitive Development: Assimilation, Accommodation, Equilibration[3]

Jean Piaget was a Swiss epistemologist whose observations of cognitive development remain today at the core of psychology's understanding of how, from infancy onward, children develop their adult ways of thinking, their formal operations of cognition.

Piaget considered himself an epistemologist first, with his primary interest being to draw conclusions about the nature of knowledge by observing how children's cognition develops.

Piaget observed that our earliest concepts are what he called topological, not metric. Our earliest concepts are those involving enclosure. We first learn about the world by learning about boundaries and relationships. It is only later in our cognitive development that we begin to focus on measurement of the concrete world.

Piaget's specific methodology through which he proposed that cognition develops is by the dual processes he called assimilation and accommodation.

Assimilation means that we take an observation or experience and add it to our existing conceptual structure, enhancing the structure that was already in place, but not transforming that structure into a new conceptual structure.

Accommodation, on the other hand, means that the observation or experience has been so novel or discordant that we cannot absorb it into our existing conceptual structure. Instead, we must modify our conceptual structure, accommodating our worldview to incorporate the latest novel, discordant event.

Piaget never heard of p-adic mathematics. But p-adic mathematics straightforwardly models Piaget's processes of assimilation and accommodation.

Assimilation means that additional elements have been introduced into what is represented by one of the digits within the prior p-adic mapping of the mind. One of these digits, which used to have, for example, up to 5 possible values, now, as a result of the most recent observation, has 7 possible values. The cognitive structure—the ordering of the p-adic digits—remains intact. But a higher p—to permit additional facts or aspects that enhance a preexisting cognitive component—is required for our enlarged understanding of an element in the train of thought.

On the other hand, accommodation, in its simplest form, means that we have one more digit to the right, one more level of hierarchy. A step was introduced that did not previously seem necessary, was not previously known. This is a new step, not incorporated within the preexisting cognitive structure, but rather modifying the preexisting structure itself.

Accomodation can also take other forms: A segment of the prior p-adic number can be preserved but encapsulated within a different enclosure. So we move the focus of our thought to a p-adically larger thought, a thought with more digits to the right.

Or accommodation can take the form of levels of enclosure collapsing into a larger single enclosure.

The point is that Piaget's model of cognition is a p-adic model. The mind interfaces with the real world, recording the world's real, metric, analog structure, and also making sense of this same world by recording its structure, using p-adic mathematics, the mathematics of enclosure.

Remember that Piaget considered himself an epistemologist first, drawing conclusions about the nature of knowledge from his observations of human cognitive development. There is another important correspondence that Piaget's epistemology gives us. This has to do with how Piaget viewed what guides the continual back-and-forth interplay between the dual processes of assimilation and accommodation.

For Piaget, this interplay is inherently unstable. An observation can trigger assimilation or it can trigger accommodation. But immediately, the cognitive structure is disrupted by another observation.

Piaget observed that the driving force in cognition is a drive for what he called equilibration, a momentary balance, a momentary resolution at a higher level, which was soon to again be disrupted, to again become unbalanced, only this time at a higher level of understanding.

For Piaget, this drive to a higher level of equilibration is the driving force not only of cognition but also of life itself. This concept corresponds remarkably to the driving principle of TGD physics, the basis of Digital Mind Math. This principle is the negentropy maximization principle, the tendency to select the route that maximizes information content (minimizes entropy).

Piaget's equilibration is Pitkänen's negentropy maximization.

Optimal Learning Is Moderately Novel

Piaget considered himself an epistemologist first, a developmental psychologist second, and an educator not at all. But his work has been used by educators to develop best practices for the educational development of the child.

One concept of Piaget's that educators have focused on is the concept that children's cognitive development proceeds best when they are exposed to situations of moderate novelty.

When the child's educational experience is too similar to her previous life experiences, she will not grow cognitively. An experience that is not novel does not offer additional details to increase the informational content of an existing idea (assimilation). Nor does it offer stimulation to restructure the organization of her idea framework (accommodation). An experience without novelty is an experience that fits perfectly within an existing cognitive structure, not offering new detail or connection or generalization.

And of course, on the other hand, some concepts will naturally be too advanced for a child at her particular level of cognitive development. Concepts that are out of the child's cognitive reach cannot offer her the bit or two of information that she can absorb into her current understanding in a way that will incrementally increase her knowledge: She cannot assimilate an observation that is too far outside her current understanding. Nor can she accommodate her cognitive structure to absorb an observation that is overly advanced—it's too much to absorb all at once.

It is moderately novel ideas that are the child's richest source for assimilation or accommodation, in the child's drive to achieve higher levels of equilibration.

These concepts—moderate novelty, and the lack thereof— offer a fascinating application of Digital Mind Math's p-adic analysis.

As the mind generates possibilities for the next thought— assimilating possibilities with higher p, and accommodating possibilities with higher negative-n (more digits to the right)—how does moderate novelty guide the mind?

The size (norm) of a p-adic idea gets larger either by assimilation (higher p) or accommodation (higher negative-n). Both processes involve novelty, new information. But which novelty is optimally novel?

We will return to this theme at a number of points in our Part Three examples. Who we are—our history of quantum jumps, as it has shaped our neural network—affects our approaches to and preferences for novelty. Perhaps too we each have an innate intellectual style, that inclines us toward small or large preferences for novelty, or toward increasing information by facts and examples, or toward increasing information by drawing broad conclusions and categorizations. Perhaps these preferences and styles change over time, as we age, as we go through different life cycles, as we learn more and become better educated.

In Part Four, we will reach deep into p-adic mathematics to find a formula that mathematically defines, within the context of Digital Mind Math, exactly what is optimally novel. This is highly speculative, since it is suggesting that there is a mathematical formula that tells us the correct way to think, to learn, to converse, to interact, to live. I don't blame you for being skeptical. But in any case, this is many pages away. Some more imaginative play first . . .

Vygotsky[4]

Lev Vygotsky was born in the Belarus territory of the Russian Empire in 1896, the same year Piaget was born. Vygotsky was aware of Piaget's work on cognitive development, and Vygotsky developed his own theories with many similarities and some differences. Vygotsky died very young, in 1934, and his work was little known in the West until several decades after his death, when his work began to become an important part of the study of cognitive development. In his lifetime, Piaget was aware of some of Vygotsky's work, but not its full scope.

A couple of the principal areas where Vygotsky diverged from Piaget are not actually central to how we are using their work in Digital Mind Math. For one thing, Vygotsky emphasized the cultural and social environment, and the role of social interaction, to a much greater extent than Piaget.

Also, Vygotsky did not accept Piaget's fairly rigidly discrete stages of cognitive develement, and instead Vygotsky viewed cognitive development as more of a continuum.

Vygotsky developed a very fluid concept analogous to Piaget's concept that optimal learning is moderately novel. Vygotsky called his concept the "zone of proximal development." Vygotsky's concept can be applied more fluidly than Piaget's, because Vygotsky was not constrained by placing the developing child within one of Piaget's four stages of cognitive development, and therefore the zone of proximal development is universally applicable without an overlay of the constraints of the child's skills being defined and limited by her sensorimotor stage, or preoperational stage, or concrete operational stage, or formal operational stage.

In case you're wondering, it is exactly this concept of zone of proximal development that we mathematicize (turn into a formula) in Part Four.

Today many educators find the zone of proximal development to be a useful concept in assessing how best to help a child learn. The term "scaffolding" is frequently used to describe the role that guided learning can play in helping the child develop.

Another aspect of Vygotsky's work that is today frequently used by educators is Vygotsky's concept of imaginative play. Here too Piaget developed a similar concept of the importance of play in the child's development, but for Vygotsky it was a more central concept.

In upcoming chapters of this book, we will revisit both Piaget's and Vygotsky's concepts:

- Piaget's processes of assimilation and accommodation are straightforwardly recorded in p-adic mathematics.
- Piaget's concept that living creatures exhibit a drive to reach

higher and higher levels of equilibration maps well to the Digital Mind Math concept that the selection of the next thought is guided by information maximization (Topological Geometrodynamics' negentropy maximization principle).

- Piaget's concept that learning proceeds best when the developing child is exposed to moderately novel concepts is, for our purposes, largely equivalent to Vygotsky's zone of proximal development, and both concepts will be used. The concept of scaffolding adopted from Vygotsky's work is particularly useful within the Digital Mind Math framework.

- Vygotsky's imaginative play will be referred to as a basic drive, a motivating force in the drive to actively construct knowledge.

10. Computers vs. the Mind

Computers today can defeat humans at chess and at Jeopardy.

These intellectual fields have in the past been put forth as examples of what humans can do best, of the special human province, of what the mind can accomplish better than computer algorithms. Yet for several decades now, computers have been better chess players than humans, and in 2011 the computer Watson defeated Jeopardy's two great human champions.

The most recent defeat for humans was the Google AlphaGo system's four-games-to-one victory over the 18-time world champion of the board game Go, which "has more possible positions that there are atoms in the universe," and which had previously been "considered impossible for computers to play at a world-class level due to the presumed level of intuition involved."[5]

Facial recognion has also long been considered a special human province, secure as a human superiority over computers for many years into the future. Yet today we're on the verge of computer superiority. There is little doubt that computer facial recognition capabilities will exceed human capability, just as they have already succeeded at chess and at Jeopardy and at Go.

Brilliant artificial intelligence specialists are needed to design and create the computer programming architecture required for these A. I. projects to succeed. But no one believes that it is solely the computer thinking processes per se that allows computers to outthink humans. The computers' advantages in these fields—chess, Jeopardy, Go, facial recognition—derives mostly or entirely from brute force, from the straightforward implications of computers being able to operate much more quickly than the human mind, and computers being able to store much more data than the human mind.

These have always been the advantages of computers—that

they can process information quickly, and that they are able to store or access large quantities of information.

What was not uniformly predicted is that these advantages—fast information processing of large quantities of information—would so soon enable computer capabilities to exceed those of the human mind, even in fields that have traditionally been considered fields of natural human advantage, fields that give the human mind advantages over computers.

What was this imagained human superiority? What aspect of chess, of Jeopardy, of Go, of facial recognition did we feel gave humans a natural advantage over computers? And what brute force process did computers use to beat us at our own human games, our games or intellectual pursuits which we feel are not simply computational, but rather require strategies, a conceptual framework, an intuition, a gestalt, understood as a whole, not built up piece by piece, mechanically, formulaically?

Many of us remain convinced that a particular and unique advantage is offered by humans' conceptual framework. Whatever this conceptual framework is, it can, for many applications, be defeated real-time by brute-force mechanical high-speed high-volume computation.

Our Last Advantage Over Computers

The human mind retains one great advantage over computers: efficiency.

Watson needed to use the energy that could power thousands of homes to defeat the human Jeopardy challengers, whose brains use only the power of a dim light bulb.

And the size of Watson's processing facility is enormous compared to the three-pound brain.

The mind must function in a way that is many orders of magnitude more efficient than computers.

That is the subject of this book: the Digital Mind Math that the human brain uses to think, with much greater efficiency than real binary mathematics allows.

Applications

Here's a prediction:

During the twenty-first century, real binary mathematics will become obsolete as the mathematics of computer science and artificial intelligence. It will be replaced by p-adic mathematics, which will take advantage of a hierarchically ordered $M^4_+ \times CP_2$ reconfiguration of space and time.

The hardware of mid-twenty-first-century computers may be biological, taking advantage of quantum biology effects,[6] emulating living systems' capabilities for a flexible-p, rather than fixed-binary, style of computation.

Or the hardware may employ the emerging technology of smart biomaterials, which are developed to respond macroscopically to small environmental changes.[7]

Or the emerging technologies of carbon nanotubes,[8] or topological qubits,[9] or of observing and recording quantum superpositions,[10] or other quantum engineering applications.[11]

This work is all underway: Brain mapping projects.[12] Postage-stamp-size chips with a million brain-inspired "neurons."[13] The neuromorphic computers and other projects of the European Union's Human Brain Project.[14]

Digital Mind Math and Topological Geometrodynamics offer an understanding of the framework and operation of the mind.

PART TWO: P-ADIC MATHEMATICS

We will now proceed to enhance our intuition for what p-adic mathematics is, how p-adic numbers are written, and what these numbers mean.

We already, from Part One, have a nice head start on this: We have discussed in Part One that the digits of a p-adic number—for example, 623.451—represent an ordered sequence of labeled objects or events. Starting from the last digit to the right, object 1 is p-adically the largest, and encloses or contains the smaller object 5, which encloses or contains the smaller object 4, and so on.

This is an excellent start, but to truly see the power of the p-adic mathematical system, we need to gain an intuition for a few more mathematical concepts.

Of course, gaining an intuition for p-adic mathematics is a bit short of taking a university course in mathematics. The purpose here is not for you to become a mathematical scholar; it is for you to gain a sense for the nature of p-adic mathematics, even though it is typically considered an advanced branch of mathematics.

It is unusual for p-adic mathematics to be introduced into an undergraduate mathematics curriculum; it is generally introduced in a graduate-level course. Even then, the full study of p-adic mathematics is an advanced mathematical specialty, studied as a full Ph.D. specialty and beyond. But here I am writing explicitly for any interested audience, even if their mathematics ended with a bit of high school algebra.

You will learn how p-adic numbers work, so that we'll be able to discuss Part Three's examples in a bit more detail, and also a bit more simply, more concisely.

One of the beauties of mathematics is the elegance with which mathematical shorthand can capture a concept, a thought, a point to be made.

Mathematicians often contrast a proof or an explanation that is accomplished by "brute force" with one that is elegant. This is not just an incidental point; it is one that is often made in criticism of some of today's great achievements of computer science.

Computers today have much faster processing speeds than the human brain, and computers can store much more data than can be stored in the brain. As a result, computers can now accomplish many tasks better and faster than humans can.

But where's the elegance in big data and superfast processors? The brain fits in our skulls—we can walk around with it. It uses a tiny amount of power, and it doesn't require industrial-size cooling systems to keep it functional.

It is the contention of this book that the brain's great efficiency derives from its p-adic structure, the structure of the elegant mathematics of enclosure, order, and organization.

And don't forget that it is a principle of this book that p-adic mathematics is our first mathematics, the mathematics with which, as infants, we begin to organize the world. This is why p-adic mathematics continues, during our whole lifetimes, to be the mathematics of thinking, the mathematics of the mind.

Although the postdoctoral papers exploring p-adic mathematics can be read and understood only by the few mathematicans who have succeeded at years of study of the subject, someone with even an elementary level of exposure—high school algebra, for example—may experience a deeply felt resonance with p-adic concepts by learning the basics of p-adic mathematics.

The hope is that, reading Part Two of this book, you will begin to internalize the mathematical system—p-adic mathematics—that is the mathematics of the mind.

Chapter 11. What Is P-Adic Mathematics? We introduce and practice the basics of p-adic mathematics, and justify the claim that there are only two complete numbering systems, real and p-adic.

Chapter 12. A Vast Informational Space. The full nature of p-adic mathematics is discussed, providing an overview for why p-adic mathematics permits our minds to efficiently work with enormous quantities of information.

11. What Is P-Adic Mathematics?

P-adic mathematics is a system of mathematics in which the numbers can look just like familiar real numbers.

For example: 123.456 can be either a real or a p-adic number.

As a real number, 123.456 is a bit larger than 100.

If we're going to proceed with any kind of accuracy, we can't lose sight that the 2 makes the number 20% larger than 100, and for most purposes, if we're talking about the real number 123.456, we wouldn't want to ignore the 3 either.

The real number 123.456 is just a little larger than the real number 123. With each digit to the right of the decimal point, we get a bit larger, and the full size of this real number is obtained only with all of its digits, down to the 6, which gives us 6 additional thousandths by which to increase the size of the number.

By contrast, p-adic numbers work entirely differently.

The p-adic number 123.456 has a totally different concept of its size.

In fact, the p-adic number 123.456 is not close in size to 100. It's close in size (the same size, actually) as the p-adic number 0.006. It's also the same size as the p-adic number 0.001.

The previous paragraph makes two different points—one point, that 0.006 and 0.001 are p-adically the same size; another point, that 123.456 is the same size as either 0.006 or 0.001.

First, let's explain the point that 0.006 and 0.001 are p-adically the same size. The p-adic numerals—1, 2, 3, 4, 5, 6, in our example— do not express size at all. They're just names, labels. So the p-adic number 1 is the same size as the p-adic number 2.

The numeral 0 still represents the absence of value. The numerals 1, 2, 3, 4, 5, 6 all represent the same-sized value.

One of the unusual features of p-adic mathematics is that the user of a p-adic number gets to label whatever she wants as 1, and whatever else she wants as 2, and so on.

As an example, let's suppose that we're using our p-adic mathematics to represent kitchen equipment. If so, we can have 1 represent the large soup pot, 2 represent the frying pan, and so on.

Doing this, there is no intention at all to indicate that the pot is larger than the pan, or vice versa. P-adic numerals are what mathematicians call ultrametric—beyond measurement, not measured. All that we're doing is creating a labeling system, in which 1 represents a pot, and 2 represents a pan. The numerals give no indication of size. They are just labels.

This ultrametric nature of p-adic numbers is one feature that distinguishes p-adic mathematics from real mathematics.

Now onto the second point from a few paragraphs back: The p-adic number 123.456 is not at all close in size to the p-adic number 100, but rather is the same size as either the p-adic number 0.006 or the p-adic number 0.001.

The way p-adic numbers work is that their size is based on the last digit to the right. It's not even based on what that digit is—since we've already established that p-adic digits are ultrametric. Nor is it based on any of the digits that appear to the left of the last digit: The p-adic numbers 123.456, 1.234, and 0.001 are all the same size. The size of a p-adic number is based on how many digits to the right of the decimal point the last digit is, no matter what that digit is, and no matter what digits precede this last digit.

And what is this size? Well, first of all we note that the more digits to the right we go, the larger a p-adic number is. This seems a bit counterintuitive, if we're used to real numbers, where we focus on the leftmost digits as the most important indicators of size. But p-adic numbers' size is based on the rightmost digit, how many digits to the right the rightmost digit is.

To be more precise about how big a p-adic number is, we need to have a talk about the numerical value of p, the number that p stands for.

Suppose that the p-adic numbers 123.456, 1.234, and 0.001 are all decimal numbers, just as we'd expect them to be if someone wrote them down for us as ordinary real numbers. If this were true— that the p-adic numbers were written in base 10—then the size (norm) of each of these three 10-adic numbers would be 1000.

A few paragraphs down, we'll explain why we have to use a number different than 10. But to keep things simple until then, let's make sure, if p were to be equal to 10, that we've mastered how the size of a p-adic number would be determined:

p^{-n}

P to the negative n.

The size of a p-adic number is found by quantifying its base, p, then seeing how many digits to the right of the decimal point the last digit to the right is (3 digits to the right, for our example), then raising the base p to that power—for our example 10 to the power 3, or 10 cubed, which is 1000.

The size of all three p-adic numbers that we're examining— 123.456, 1.234, 0.001—is 1000, if the base p were p=10, because all three of these numbers have three digits to the right of the decimal point.

The reason that the formula above for the size of a p-adic number had to say p^{-n}, rather than p^{n}, is due to how we count the digit numbers, how we count the place number at which each digit resides. For example, for the number 123.456:

1 is in place number n=2
2 is in place number n=1
3 is in place number n=0
4 is in place number n=-1
5 is in place number n=-2
6 is in place number n=-3

As you can see, we label the place number of digits to the left of the decimal point with positive numbers such as n=2 or n=1 (n=0 for the final digit to the left of the decimal point). And we label the place number of digits to the right of the decimal point with negative numbers. So the thrid digit to the right of the decimal point is digit number n=-3, and the size of a p-adic number whose last digit is three digits to the right of the decimal point is p raised to the power of negative negative 3, or p to the power of 3 (that is postive 3, since two negatives make a positive). Usually, we'd pronounce this p cubed, rather than p to the power of three, but they mean the same thing, and except for p to the power of 2 (p squared), there aren't shorthand terms like squared or cubed, and we have to say p to the fourth power, or p to the 100th power, or p to the power of 100.

To master this, let's try some more examples. For simplicity of calculation, for all of our examples we'll assume that the base of these p-adic numbers is base 10. Again, this is an arithmetic simplification that very soon we'll see is an oversimplification, an impossibility actually. We'll get right to this after we master the calculation of the size of p-adic numbers—the norm of p-adic numbers, to use the mathematician's term.

For all these examples, assume that the base p that is used for the p-adic representation is base p=10:

The size (or norm) of the 10-adic number (p-adic number based on a base p of p=10) 123.45 is 100.

The size of 12.34 is also 100.

The size is also 100 for the numbers 123456.01, 6.06, 0.12, 0.05, and 0.03.

The size of 12345.6, or 123.4, or 1.2, or 0.6, is 10.

Now let's stretch ourselves a bit by continuing to make the numbers p-adically smaller, by continuing to make the final digit to the right one fewer digit to the right. We'll proceed this way in order to see how this continues to work even when there are no digits to the right of the decimal point, and the final nonzero digit on the right gets further and further to the left.

Still assuming that p=10 for all of these examples of p-adic numbers, let's confirm how this progression works:

The size (or norm) of the p-adic number 123.456 is 1000.

The size of either 123.45, or 234.56, or 1.23, is 100.

The size of 123.4, or 234.5, or 6.2, is 10.

The size of 123, or 456, or 25, or 2, is 1.

The size of 120, or 450, or 20, or 10, or 60, is 0.1 (that is, one tenth, also written 10^{-1}).

The size of 123,400, or 500, is 0.01 (one hundredth, or 10^{-2}).

The size of 123,000, or 4,000, is 0.001 (one thousandth, or 10^{-3}).

The size of 1,000,000 (one million), or 1,234,000,000(1 billion, 234 million) is 0.000001 (one millionth, or 10^{-6}).

Perhaps at this point you're convinced of the elegance of the mathematician's shorthand: The size, or the norm, of a p-adic number is p^{-n}, or for our examples above 10^{-n}, where n is the place number of the last digit to the right:

The norm (size) of the p-adic (in this case, 10-adic) number 123.456 is $10^{-(-3)} = 10^{+3} = 10^3 = 1000$.

The norm of 123.45 is $10^{-(-2)} = 10^{+2} = 10^2 = 100$.

The norm of 123.4 is $10^{-(-1)} = 10^1 = 10$.

The norm of 123 is $10^0 = 1$. (This might stretch the mathematical recollection of some: Any number raised to the zero power is 1.)

The norm of 120 is $10^{-1} = 0.1$. (Note that, to determine the norm, we look at the last nonzero digit, ignoring the following zeroes.)

And so on. You see the pattern: The norm (size) of a p-adic number is p^{-n}, where n is theplace number of the last digit to the right (place numbers being positive to the left of the decimal point, ending with place number 0 for the final digit to the left of the decimal point, then continuing as negative place numbers for digits to the right of the decimal point).

One final repetitive paraphrasing, then we'll move on: Progressing from n=-2 when there are two digits to the right of the decimal point, and n=-1 when there is just one digit to the right of the decimal point, we've seen that n=0 when the last digit is just to the left of the decimal point, n=+1 when the last digit is in the 10s position, n=+2 when the last digit is in the 100s position, and so on.

Enough of this. It is time to see why p=10 in our examples above was for simplicity of calculation, but is actually an impossibility: P stands for prime number, and p-adic numbers can be based only on prime numbers, not on a number such as 10, which is not a prime number, since $10 = 2 \times 5$ and is therefore not a prime number.

Then, after a little more experience with the mathematician's use of the term norm, we'll be ready to dig more deeply into p-adic numbers, in order to discuss: Why bother with p-adic numbers? Why are they important? Why are they useful?

But first: prime numbers.

Prime Numbers Only

The p in p-adic numbers stands for prime numbers. P-adic numbers make sense only when the digits in the p-adic number have a base that is a prime number.

Power series and summations. The real numbers that we're used to working with have a base of 10. That is, they're decimal numbers. We can surmise that we're comfortable with these numbers because we have 10 fingers, but for whatever reason decimal numbers seem easiest for us.

In decimal numbers, a single digit, like 9 or3, stands alone with its own value, but technically this is the unit's digit, representing how many 1s (units) we have, or, said another way, how many 10^0s we have (10 to the power zero is 1, since any number to the power zero is 1).

The digit immediately to the left of the unit's digit—the 2 in 20, or the 7 in 70—represents how many 10s we have (technically, how many 10^1s we have, how many 10s to the first power, since any number to the first power is the number itself).

Then, moving further to the left—the 5 in 500, or the 3 in 300—represents how many 100s we have (how many 10^2s, how many 10 squareds).

And if there are numbers to the right of the decimal point, this is where negative exponents come in, negative powers of 10. For example, 0.4 means there are are 4 tenths, 4 10^{-1}s, 4 10s to the power negative 1.

And 0.08 means there are 8 hundredths, 8 10^{-2}s, 8 10s to the power negative 2.

Mathematicians often use what is called a power series to break up a number—for example, the real base-10 number 2987.65—into its components as powers of 10:

$$2987.65 = 2*10^3 + 9*10^2 + 8*10^1 + 7*10^0 + 6*10^{-1} + 5*10^{-2},$$
where the * indicates multiplication.

However, this form of the power series is a bit tedious, and not elegant enough for most mathematicians. There is a lack of aesthetic appeal to all those repeating powers of 10 that are being added together.

So typically mathematicians represent the above power series using the Greek letter Σ (capital sigma), where the sigma Σ stands for sum (addition):

$$\sum_{i=-2}^{3} a_i \, 10^i$$

This is read: The sum, from i=-2 to i=3, of a-sub-i times 10 to the i.

The symbol a_i is read as "a-sub-i," and for our example 2987.65:

$a_{-2} = 5$

$a_{-1} = 6$

$a_0 = 7$

$a_1 = 8$

$a_2 = 9$

$a_3 = 2$

Note that, in its elegance, if no mathematical operator is indicated—no plus sign or minus sign or multiplication sign or division sign—multiplication is assumed, which is why we don't need a * or a × between a_i and 10^i.

And what this elegant shorthand means is that we're creating a sum that has six terms in it, each term structured similarly, with the first term starting by using the number -2 for the symbol i, the second term using -1 for the symbol i, the third term using 0 for the symbol i, the fourth term using 1 for the symbol i, the fifth term using the number 2 for the symbol i, and finally the sixth term using the number 3 for the symbol i.

Note that typically these Σ summations are constructed reading from the lowest number (-2 in this case) underneath the Σ, to the highest number (3 in this case) above the Σ, even though we're used to writing numbers from left to right in the opposite order. Technically speaking, because addition is commutative, it doesn't matter in which order we write the summation—it's just a more typical convention that summations are indicated starting from the lowest power to the highest power.

So the above summation, written concisely using the Greek

letter Σ, can be expanded into its power series as:

$$a_{-2} \, 10^{-2} + a_{-1} \, 10^{-1} + a_0 \, 10^0 + a_1 \, 10^1 + a_2 \, 10^2 + a_3 \, 10^3$$

By the way, the subscript 3 in a_3 is a totally different animal from the superscript 3 in 10^3. The superscript is an actual mathematical operator, indicating that we raise 10 to the power 3, so that $10^3 = 10$ cubed $= 1000$. But the subscript 3 in a_3 is just a way to indicate that the coefficient that multiplies 10^3 is this particular term named a_3 of the power expansion. With our example being the real decimal number 2987.65, this means that $a_3=2$, since 2 is the coefficient of the 10^3 term in the power expansion.

In other words, substituting the actual digits a_i within our expansion of the power series for the example 2987.65, the summation would be expanded as:

$$5*10^{-2} + 6*10^{-1} + 7*10^0 + 8*10^1 + 9*10^2 + 2*10^3$$

You can see the temptation of using the more concise

$$\sum_{i=-2}^{3} a_i \, 10^i$$

even if it does use Greek letters. It's elegant. Not a brute-force expansion written as the addition of six terms.

Now you can certainly at this point argue that 2987.65 is also quite elegant. And you're correct. However, we have our reasons for mastering the Σ-based notation system for summations: We will soon be leaving the world of base 10, decimal numbers. In fact, we will be going to a world very far away: 11-adic numbers that can be written to look just like decimal numbers are just the beginning. Only the Σ-based notation can be used on the very foreign planet where we will soon arrive.

By the way, sometimes this Σ summation is written more linearly as

$$\sum_{i=-2}^{3} a_i \, 10^i$$

which means the same thing. I mention this because it leads to some more elegant mathematical symbology, which probably stretches the limits of our all-you-need-is-high-school-algebra rule, but I'll introduce it now because it will come in handy later on.

In an effort to make as general as possible our summation of the power series expansion for real decimal numbers, we do not want to limit the summation to only the particular i's ranging from -2 to 3. If we're trying to make this as general as possible, we need to start and end at any negative or positive whole number (that is, at any integer).

Close your eyes and turn the page if this is getting too technical for you, but it will be useful to gain some familiarity with a couple more symbols, starting with the symbol $\mathbb{Z}$, a fancier, double struck capital letter Z, which is the mathematician's symbol for the set of integers (postive whole numbers plus 0 plus negative whole numbers).

The general form of the power series expansion for real decimal numbers is therefore:

$$\sum_{i \in \mathbb{Z}} a_i \, 10^i \text{, where } a_i \in \{0, 1, \ldots, 9\}$$

Here (our last symbol for now) the symbol $\in$ is read "is an element of," so that the above expansion is read: The sum, for all integers i, of the product a-sub-i times 10 to the i, where each coefficient a-sub-i can take on a value from the set of whole numbers 0 through 9.

This is the most general form that a real decimal number can take—a sum consisting of a series of coefficients (whole numbers from 0 through 9) multiplied by an integer (positive or negative whole number or zero) power of 10.

Whew.

This is a lot of mathematical terminology for something you already know—what a decimal number is.

But it's easier to get used to this terminology with something you're already familiar with, so that we can apply it in the less familiar sphere of p-adic numbers.

As an aside before we move on, let's think about the point that you already know how to write a decimal number, whether or not you know about Σ or $\in$ or i. Now you know how your mind feels: Your mind knows how to use p-adic numbers, without knowing about the exotic mathematical concepts that we'll soon be discussing.

Only p-adic numbers are worthwhile; g-adic numbers are too troublesome. We've already discussed at length how a p-adic number can be written just like a real number: The same type of power series expansion, elegantly summarized with our Greek letter sigma Σ summation, can be used to represent either a real decimal number or a p-adic number.

The only difference is that real decimal numbers use base 10, whereas p-adic numbers use as their base any prime number p.

So p-adic numbers can look just like real decimal numbers, even though their meanings are quite different, even though their size (norm) is quite different.

Before moving on, let's discuss a little more what we mean by a p-adic number using as its base any prime number p: Why can't we use nonprime numbers as the base, such as base 10, or for that matter, base 4, or base 6, or base 12? And what do we mean by a p-adic number using any prime number p as its base, rather than just a particular prime number such as 7, or 2, or 3, or 11?

With respect to why only prime numbers form useful bases, let's first note that a mathematical system can be constructed, using p^{-n} as its norm, but not restricting the base to only prime numbers p, instead permitting any whole number g as its base. Some development of such a mathematical system has been studied—these are called g-adic numbers—but the g-adic mathematical system is troublesome for most practical purposes.

To see why, recall that, in our ordinary real mathematics, we sometimes run into the problem that division by zero is undefined. If we face a situation in which we're dividing by zero, we've got an unresolvable problem and can't go on. We need to rethink how we're proceeding, or we need to set up special rules. But there is no direct solution to how to divide by zero: Division by zero is undefined and therefore impossible.

This is what happens over and over when we try to construct a mathematical system (g-adic mathematics) in which we permit the base of the system to be any whole number g, rather than limiting our base to prime numbers p. In g-adic mathematics, whenever we let g be a nonprime number, we find frequent division by zero, frequent undefined operations, which require workarounds and special rules. So the most extensive mathematical development has been for p-adic mathematics, limiting the base to prime numbers p, and it is this p-adic mathematics that is the basis for Digital Mind Math.

As to the question of permitting any prime number as the value of p, this is analogous to what we are able to do with real mathematics also. After all, binary mathematics (base 2, where the digits can be only 0 or 1) is a common system of mathematics used in computer science and other fields. The basic difference is that p-adic mathematics is almost always developed generally, so that it can be used for any prime number p. This means that to be specific, we need to disclose what value of p we're assuming. This is actually true for real mathematics also; it's just that it's so common to use base 10 (decimal numbers) that we're not used to thinking about it. It's something of an exception to use binary numbers, so we're quite aware that, if that's what we're using, we need to say so.

While we're on the subject of decimal numbers, one final point before we go on. We've been a bit sloppy talking about the decimal point when we use the point for p-adic numbers. After all, words have meaning, and decimal points are for decimal numbers (base 10, not base p). However, the term decimal point is so familiar, I've elected in this book to use it even for p-adic numbers, to make the discussion a bit more intuitively appealing, even though it's not strictly correct.

Norms

We've established that the mathematical concept of norm is essentially equivalent to size: How does a mathematical system conceptualize and measure size?

For real numbers, the norm (the size) is the absolute value. This is a pretty straightforward concept, in which a postive number, such as 123.456, has as its absolute value 123.456. And a negative number, such as -123.456, also has 123.456 as its absolute value.

We've also commented on a couple of elements of the real norm, the absolute value, that are entirely familiar to us, but, perhaps oddly, are not universal elements of norms. In particular, we noted that the most important number in defining the real norm is the first number on the left, whereas the most important number (in fact the only important number) in defining the p-adic norm is the last number on the right. This leads to a second important and perhaps surprising distinction between the real norm and the p-adic norm: For the real norm (absolute value) all the digits matter, but for the p-adic norm only the last digit on the right matters. And, for the p-adic norm, the value of this last digit doesn't matter at all; all that matters is where in the p-adic number's expansion it appears (how many digits to the left or the right of the decimal point).

In any case, we've already established that there are different ways for mathematical systems to conceptualize size, to define the norm. What we haven't explicitly established is the pretty amazing fact that the entire number system derives from how the norm is defined.

If we set up some pretty basic conditions for how we expect addition and multipliction to operate (these conditions are, mathematically, that the mathematical system is what is called a field), it is the norm that then defines the entire mathematical system. We can define the norm as the absolute value, and we get the field of real numbers. Or we can define the norm as p^{-n}, and we get the field of p-adic numbers.

This is surprising in and of itself, that the definition of the norm can so powerfully determine the entire mathematical system.

But much more surprising is that there are only two reasonable ways to define the norm.

There Are Only Two Reasonable Complete Mathematical Systems, Real and P-Adic

It seems reasonable enough to require a mathematical system to operate as a field, that is, a mathematical system with addition and multiplication, including the numbers 0 and 1, and operating according to our normally expected commutative, associative, and distributive laws.

If we add just one more condition—that the number system be a complete number system—it turns out that there are only two possible number systems, real numbers and p-adic numbers, that meet what appear to be entirely reasonable conditions.

What's surprising is that only two systems meet these conditions, since the conditions seem so broad and nonrestrictive.

What we mean by a complete number system is that it fills in all of the holes between numbers. Pretty obviously, the set of integers is not a complete number system, since integers are limited to positive and negative whole numbers plus zero. Clearly we need fractions in between the whole numbers. Otherwise the number system would not be complete.

The broadest definition of a number system that includes fractions is called the set of rational numbers. Here the word rational derives from ratio, and the set of rational numbers is the set of all numbers that can be constructed as ratios of integers. This means that all decimal numbers are included, since any decimal number, such as 284.8905, can be constructed as a ratio of whole numbers, in this case the ratio 2,848,905 divided by 10,000.

It's tempting to think that rational numbers form a complete set, since we can have lots and lots of digits to the right of the decimal point, but it is not true that this fills in all of the holes between numbers. A popular counterexample is the number square root of 2. The square root of 2 is not a rational number. It cannot be constructed as the ratio of any two whole numbers: No matter how many numbers we extend to the right of the decimal point we will never get exactly the square root of 2. Said another way, the square root of 2 is a real number, but it is not a rational number.

Once we've bitten the apple and found that there are real numbers that are not rational, we may start to wonder if real numbers form a complete set, if real numbers fill in all the holes between numbers.

Here's where Ostrowski's Theorem comes in. This theorem, developed in the early twentieth century by the European mathematician Alexander Ostrowski, states that there are two and only two norms that complete the rational numbers, that fill in all of the holes between numbers, so that the set of numbers is complete. One of these norms is the absolute value, which results in (induces) the field of real numbers, and the other is the norm p^{-n}, which induces the field of p-adic numbers.

This gets p-adic numbers into the finals for the best numerical system for the mind to use in thinking, for the most evolutionarily promising way for our brains to have adapted in order to efficiently think.

In fact, it may be apparent at this point that p-adic mathematics is better at mind mathematics than real mathematics is, due to its intuitive appeal as a method for thinking : p-adic mathematics' inherent labeling and categorization possibilities, rather than real mathematics' metricization.

We have discussed in Part One that the digits of a p-adic number—for example, 623.451—represent an ordered sequence of labeled objects or events: Starting from the last digit to the right, object 1 is the largest, and encloses or contains the smaller object 5, which encloses or contains the smaller object 4, and so on.

But the real trump card that explains p-adic mathematics' evolutionary advantage in creating the brain's thought processes is the vast informational space in which p-adic mathematics operates, the vast amount of information that a single p-adic number represents.

This will be discussed in the following chapter.

One More Simplification: Renormalization and Scaling

There are a number of flexibilities that make p-adic numbers easy to use. You may not yet be convinced that it's appropriate to use "p-adic" and "easy" in the same sentence. But give it time.

In Digital Mind Math, our basic use of p-adic mathematics is during two continually repeated processes of thinking: We generate possibilities for the next thought by extrapolating from the p-adic structuring of the most recent thought. And we compare the sizes of these possible next thoughts by calculating their p-adic norms.

This, leading to the real experience of the selected next thought, is the basic cognitive process, which recurs frequently and quickly. This gives us several different possibilities for describing the process more understandably.

For example: We can start this core cognitive process with some arbitrarily scaled p-adic decimal. This seems to make many of our upcoming examples easier to understand: We can look at possible next thoughts that add levels of detail by adding digits to the right. Or we can add details at the same level by increasing the maximum size of the p-adic digits. Or we can branch to other thoughts that partially match the digits of the current thought.

Alternatively, we can start the core cognitive process fresh each time with a p-adic integer (whole number, with no digits to the right of the decimal point). Then, at the possibility-generation phase of the repeated core cognitive process, we vary all the digits in many ways—change digits, add digits, move to the left, move to the right. During this possibility-generation phase, some of the possible next thoughts will be labeled with p-adic decimals, with digits to the right of the decimal point, whereas other possibilities will be labeled as p-adic whole numbers with trailing zeroes. If it makes the discussion easier, we can, after experiencing the selected p-adically numbered thought and preparing to start our possibility generating again, use the analytic convention that (although we'll keep all the digits in order) we'll move the digits so that we start the next cognitive cycle with a p-adic integer filled in all the way to the units digit (that is, a p-adic integer with norm of 1).

As long as we keep the digits in order, this renormalization or scaling will be fine for discussion of Digital Mind Math's core cognitive process, because of the cyclical nature of the process, starting each cycle by generating possible next steps to follow the prior experienced thought. Therefore, in our discussions ahead, we will be using various patterns of p-adic decimals and p-adic integers, depending on what seems to make the intuitive discussion more accessible.

Said another way, this renormalization or scaling works just fine because it sacrifices nothing in terms of our ability to generate posibilities for our next thought, and because the selection process — calculating the p-adic norm in order to identify the information-maximizing thought to experience — is a process that compares the possible next norms one to another. So why not keep things simple and start each cognitive cycle with a norm of 1, which is what we start with when we have a p-adic integer, with a nonzero digit in the first position to the left of the decimal point, and no digits to the right of the decimal point.

Mathematicians take this renormalization or scaling process for granted, as a triviality. In fact, later in *Digital Mind Math* we will see that some important theorems have been developed using p-adic whole numbers only — no p-adic decimals. In this situation — never allowing p-adic decimals — we'll want to create our intuition by playing around with p-adic numbers like 1234000000 and 1234500000 and 1234560000, to give us flexibility about where the locator n is of the last digit to the right.

Yes, it remains true that p-adic numbers are bigger when their last digit is further to the right. But we will find that very often our interests are in comparing one p-adic number to possibilities for a next p-adic number, and in comparing these possibilities to each other. So very often we have the flexibility to continually reset — renormalize, or rescale — our starting point in order to help us with our intuition. In these cases, we're on firm ground with these renormalizations as long as we keep the digits in order, even if we change where the decimal point is or where the last digit is to the right.

12. A Vast Informational Space

Ostrowski tells us that there are only two complete numbering systems—real and p-adic—two systems of mathematics that fill in all of the holes between numbers.

But, when we think more about what it would take to create an ideal numerical system, there's one more characteristic that we'd like it to exhibit: We'd like the mathematical system to be closed: We'd like any equation written in that mathematical system to have a solution in that same mathematical system.

This seems pretty obvious, perhaps trivial—but it's not as trivial as it might seem at first glance.

Closing the Set of Real Numbers

Early man and woman, sitting by the campfire, trying to avoid getting eaten by mammoths, enjoyed taking their minds off their problems by solving mathematical equations.

At the time, they only knew how to count—1, 2, 3, 4, . . .— using the natural numbers, also known as the counting numbers.

This worked great for a long time: $x + 4 = 7$ had the solution $x = 3$. $x + 984 = 999$ had the solution $x = 15$.

It looked for a long time that the counting numbers or natural numbers formed a closed set: It looked like any equation using counting numbers had a solution that was also a counting number. Until one day someone came up with the equation $x + 172 = 172$. This did not have a solution that is a counting number. The counting numbers do not form a closed mathematical set, because we can write an equation using only counting numbers that does not have a solution that is a counting number.

The solution was called 0, and we now had the set of whole numbers, or nonnegative integers, which consisted of the counting numbers (or natural numbers) plus the number 0. But it wasn't too long before someone came up with the equation x + 7 = 5. This resulted, of course, in the idea of adding negative integers to our set of whole numbers (nonnegative integers), forming the full set of integers (positve, negative, and 0). But, as it turned out, still not closed.

After all, the equation x × 2 = 1 is written using only integers, but it does not have a solution that is an integer. Thus fractions, and the entire set of rational numbers—numbers that are the ratio of two integers—were developed. Mathematicians use the double-struck capital letter $\mathbb{Q}$ as the symbol for the set of rational numbers. (Preview: They can't use the double-struck capiltal letter $\mathbb{R}$ to represent the set of rational numbers, because there is a more important use of $\mathbb{R}$ ahead.)

But the set of rational numbers is still not closed. After all, the simple equation $x^2 = 2$ has as its solution $x = \sqrt{2}$ (the square root of 2), which is not a rational number. There is no ratio of any two integers that exactly equals the square root of 2. So we needed to expand our set of numbers even further, to the set of real numbers $\mathbb{R}$.

Now it was tempting to think that this would do the trick. After all, it was this very expansion—from the set of rational numbers to the set of real numbers—that we have already shown created a complete number system. Wouldn't it be nice if this also created a closed number system?

No such luck.

The simple equation $x^2 = -1$, using only real numbers, does not have a solution that is a real number. So we invented a new symbol, i, to represent the solution to this equation. Here i stands for imaginary, and when we add the set of all imaginary numbers (all multiples of i) to the set of real numbers, this set is called the complex numbers $\mathbb{C}$.

It turns out that the set of complex numbers is a closed set. All equations written in complex numbers have a solution that is a complex number.

It also turns out that the set of complex numbers remains a complete set, with no holes in it. This may seem obvious, since it was built up from the set of real numbers, which form a complete set, but it's not obvious. It needs to be proven, and it has been proven. (Spoiler alert: When we go through this analogous process for p-adic numbers—closing the complete set—the resulting set will no longer be complete!)

So it turns out to be fairly straightforward to start with a complete set of real numbers, and expand it to the closed complete set of complex numbers: Starting with the real numbers, all it takes is to add the imaginary number i, which was found as the solution to the quadratic (second-order) equation, $x^2 = -1$, and we produce one of mathematics' only two closed complete numbering systems.

As an aside, imaginary numbers could have used a good publicist so that they wouldn't have been tarnished with the name imaginary. They're actually as real as real numbers, no more imaginary than real numbers, but the problem was that the term real had already been taken. This problem was compounded by labeling as complex numbers the union of the set of real numbers and the set of imaginary numbers. Complex numbers—the name alone has turned off generations of potential math students.

Closing the Set of P-Adic Numbers

We've already discussed how Ostrowski's Theorem tells us that there are only two ways to complete the set of rational numbers $\mathbb{Q}$. One completion of the set of rational numbers $\mathbb{Q}$ is the set of real numbers $\mathbb{R}$. The other completion of the set rational numbers $\mathbb{Q}$ is a basic set of p-adic numbers that mathematicians call $\mathbb{Q}_p$.

Right away, this labeling foreshadows some problems with this p-adic completion $\mathbb{Q}_p$ of the set of rational numbers $\mathbb{Q}$. This set of numbers, although larger than the set of rational numbers $\mathbb{Q}$ that we're completing, must not be important enough to get its own letter.

$\mathbb{Q}_p$ is complete, just as $\mathbb{R}$ is complete. Both $\mathbb{Q}_p$ and $\mathbb{R}$ complete the rational numbers, fill in all of the holes between rational numbers. In fact, according to Ostrowski, they are the only two ways to complete the rational numbers.

But just as the complete set of real numbers $\mathbb{R}$ is not closed, so too is the complete set $\mathbb{Q}_p$ of p-adic numbers not closed. We can write equations in $\mathbb{Q}_p$ that do not have a solution in $\mathbb{Q}_p$.

In our attempt to enlarge $\mathbb{Q}_p$ in order to close it, it turns out that we need a much greater enlargement than the enlargement needed to close the real numbers by adding the imaginary numbers. You will recall that the imaginary numbers derived from a second-order equation written using real numbers. But closing $\mathbb{Q}_p$ requires an infinite-order equation, so the set produced, called $\overline{\mathbb{Q}}_p$, more elaborately extends $\mathbb{Q}_p$ than $\mathbb{C}$ extends $\mathbb{R}$.

Moreover, as we've already mentioned above as our spoiler, the set $\overline{\mathbb{Q}}_p$, created by enlarging $\mathbb{Q}_p$ in order to close it, is no longer complete. So we need to expand this set even further in order to complete it.

It turns out that this expansion needs to be done in two steps—first a p-adic completion to the partially complete set called $\mathbb{C}_p$, then a spherical completion to the set Ω_p, using the last letter of the Greek alphabet, omega, which also finally gives us our closed complete extension of the p-adic numbers.

Sometimes this closed complete set of p-adic numbers is simply labeled Ω, without the subscript p, which is a little sloppy (because everything about p-adic mathematics is specific to the specific value of p) but easier to write.

The set Ω is an enormous mathematical space, many orders of magnitude larger than the set of complex numbers that is the only other closed completion of the rational numbers:

- It takes an infinite-order equation, rather than real numbers' quadratic (second-order) equation, to close the set of p-adic numbers.
- Unlike the situation for complex numbers, the set of p-adic numbers that has been expanded for closure is now no longer complete.
- There are two further steps of expansion in order to finally create the closed and fully complete numerical space Ω in which p-adic number live.

What this means is that each p-adic number, as fully represented in the numerical space Ω, represents an enormous quantity of information, far beyond what real numbers or even complex numbers have accustomed us to.

Accessing the Vast Numerical Space Omega Ω: Introduction to Teichmüller Representatives

We have now reached an area of highly advanced mathematics, so this will be just an intuitive introduction.

You will recall that the p-adic numbers 1, 2, 3, . . . are ultrametric, that is, just labels, and not indicative of size. The p-adic number 1 is not bigger or smaller than 2 or 3; they are just labels and are all the same size.

Nevertheless, even though it doesn't seem to make any difference what we call these labels, it turns out that one particular set of p-adic numerals permits the full realization of p-adic mathematics in the great mathematical space Ω. These particular p-adic numerals are called Teichmüller representatives, after their mid-twentieth-century developer, the German mathematician Oswald Teichmüller.

We won't do anything highly technical with this, but we will proceed with a brief introduction.

For any value of p, the number of nonzero Teichmüller representatives is p-1. For example, in 7-adic mathematics, there are a total of seven Teichmüller representatives: the number 0, plus six nonzero numbers.

The p-1 nonzero Teichmüller representatives all have different values, but the value of each of them is a (p-1)th root of 1.

For example, if p = 7 (that is, we are using 7-adic mathematics), it turns out that, there are six different 7-adic numbers, which, when multiplied by themseleves six times using 7-adic mathematics, equal 1.

It turns out to always be true, for any value of p, that there are p-1 distinct (p-1)th roots of 1, and that using these roots is key to allowing p-adic numbers to access the full scope of information that Ω-space represents. The basic reasons for this are that: (1) Teichmüller representatives give us p-1 different names for the p-adic digits (for example, 6 different 7-adic 6th roots of 1), which, when combined with 0, always give us p different digits; (2) they simplify p-adic multiplication; and (3) there are different Teichmüller representatives for different values of p, which is not true if we used the digits 1, 2, 3, . . .

The p-adic concept that there are always p-1 nonzero roots of 1 can be summarized using the equation:

$$x^p - x = 0.$$

This equation always has p roots (p solutions): the number 0, plus the p-1 (p-1)th roots of 1.

This is starting to get to be a mouthful, but it is worth gaining a bit of an intuition for two mathematical miracles of sorts, one that we just established, and one that we're about to discuss.

The mathematical miracle that we've already mentioned is

that there are always p-1 distinct (p-1)th roots of 1 for a given p-adic p. This is important, because it's what allows us to proceed with Teichmüller representatives instead of the digits 0, 1, 2, 3, . . .

How can theseTeichmüller representatives all be distinct (different from each other, not the same value), yet on the other hand each be described as a root of 1?

Something of an intuitive appeal for this concept can be gained by going back to our ordinary complex numbers (that is, real numbers plus imaginary numbers). Let's examine the concept that the number 16 has four distinct complex roots.

We know that $2 \times 2 \times 2 \times 2 = 16$. So 2 is one of the four fourth roots of 16.

Also, $(-2) \times (-2) \times (-2) \times (-2) = 16$, taking advantage of the the fact that, when two negative numbers are multiplied, the product is a positive number. So -2 is the second of the four fourth roots of 16.

The third of the four fourth roots of 16 uses the imaginary number i, which is $\sqrt{-1}$, the square root of -1. 2i is also a fourth root of 16, because $2i \times 2i = 4i^2$, which is -4, since $i^2 = -1$. So $(2i)^4 = (2i \times 2i) \times (2i \times 2i) = (-4) \times (-4) = 16$.

Finally the fourth of the four fourth roots of 16 is -2i. This can be seen by first noting that $(-2i) \times (-2i) = 4i^2 = -4$. So $(-2i) \times (-2i) \times (-2i) \times (-2i) = (-4) \times (-4) = 16$.

So this gives us the concept that the real number 16 has four distinct fourth roots: Two are the real numbers 2 and -2; the other two are the imaginary numbers 2i and -2i.

With this concrete example in the sphere of the more familiar real and complex numbers, this gives us a bit of grounding in—and respect for the mathematically miraculous nature of—the concept that, in p-adic mathematics, for any value of p, there are always p-1 (p-1)th roots of 1. In 13-adic mathematics, there are twelvedistinct twelfth roots of 1. And in 23-adic mathematics, there are twenty-two distinct twenty-second roots of 1. These, along with the number 0, are the Teichmüller representatives.

The second mathematical miracle is that there is an amazing correspondence between the nonzero Teichmüller representatives and the ordinary counting numbers 1, 2, 3, . . .

For example, if we are talking about 7-adic mathematics, you can start with one of the six nonzero Teichmüller representatives, and keep adding or subtracting 7 p-adically, and you will get one of the numbers 1, 2, 3, 4, 5, or 6. Do the same for another of the six nonzero Teichmüller representatives, and you will get a different number 1, 2, 3, 4, 5, or 6. And so on: There is a direct one-to-one arithmetic correspondence between the six nonzero Teichmüller representatives and the six ordinary whole numbers 1, 2, 3, 4, 5, and 6.

This is summarized mathematically, for each of the Teichmüller representatives a_j, and for each of the whole numbers j (0, 1, 2, 3, . . . , p-1), as:.

$$a_j \equiv j \ (\mathrm{mod}\ p)$$

Here, the equation above means that each Teichmüller representative a_j is "congruent modulo p" ($\equiv$ mod p, which means equal after subtracting or adding p or multiples of p) to one of the whole numbers 0, 1, 2, 3, . . . , p-1.

This is a little hard to see mathematically, because we won't be going into the details of p-adic arithmetic for fractions and negative numbers. But the points to keep in mind are that there are always p distinct Teichmüller representatives (0 plus the p-1 (p-1)th roots of 1—the first mathematical miracle), and (the second mathematical miracle) that each of these p-1 nonzero Teichmüller representatives is intimately connected to one of the simple counting number 1, 2, 3, . . . , p-1.

So we really are left wondering: What do mathematicians obtain from using Teichmüller representatives rather than ordinary counting numbers?

Accessing the Vast Numerical Space Omega Ω: P-Adic Numbers in Their Full Splendor

A fully realized p-adic number is represented by the power series

$\sum_{r \in Q} a_r \, p^r$, where a_r is a Teichmüller representative for the particular value of p.[15]

Let's examine this summation (this power series) for a moment, because we've snuck something else in that's quite amazing, besides Teichmüller representatives.

We've already established that, for reasons that we've only hinted at and won't be going into in detail about, the coefficients of these fully realized p-adic numbers—operating in the vast informational space Ω—must be these strange root-of-1 Teichmüller representatives.

But look further, at the index for the summation not simply being elements of the set $\mathbb{Z}$ of integers (. . . , -3, -2, -1, 0, 1, 2, 3, . . .). The digits of these fully realized p-adic numbers do not appear one for each integer. They appear one for each rational number (each decimal; each fraction of any kind), as many rational numbers as you want, with any number of decimal points!

In other words, in order for a single p-adic number to represent what it stands for in the vast informational space Ω, this single p-adic number consists of a sum of a whole sequence of digits (in the form of Teichmüller representatives), each multiplied by an infinitely fractionable power of p.

Suppose you have a p-adic number describing your understanding of a thought or an event or a concept. Then you come across a refinement that introduces an extra detail or extra step. You do not need to throw out the original p-adic number, or change all the powers of p. Instead, you may simply insert a term using a rational power of p in between two terms that are already there.

A fact or a situation or a story or a thought resembles one previously considered, but more layers of information are now known? Adjust the previous thought wherever it's needed by simply inserting the new information.

Witt Vectors

Teichmüller representatives permit the creation of a rich mathematical system for calculations using p-adic numbers. They do this by representing p-adic numbers as sequences (ordered lists) of digits called Witt vectors.

Witt vectors were developed in the mid-twentieth century by the German mathematician Ernest Witt and are well beyond the scope of this book. One mathematician working with them, Hendrick Lenstra, has said about them: "The formulas do not fit in the head of a civilized mathematician of the twenty-first century."[16] But as an appetite whetter I offer the concept that it is these Witt vectors that permit—through structure-preserving maps such as those called the Frobenius lift or Frobenius endomorphism, the Verschiebung (or shift) morphism or shift operator, and the restriction map—the orderly manipulation of sets of p-adic numbers.

So a p-adic number, in Digital Mind Math, represents an ordered sequence of events. Events can be labeled with unlimited detail and can be sequenced broadly or in a highly refined manner, with as many layers of detail as needed. It is easy to further refine the sequence of events, by sticking a rational power of p in between the powers that have already been contemplated. This is, so far, all one single p-adic number.

These p-adic numbers can then be grouped together and

examined and manipulated, as a "ring" of Witt vectors—a set of Witt vectors with a system of rules for manipulating them.

This is the vast informational space of p-adic numbers.

But we still have two more levels of vastness!

Universal Witt Vectors

Much of p-adic mathematics is developed under the assumption that there is a specific value for p. While it is true that at different times p may take on the value of any prime number, under this "p-typical" approach—where p takes on a specific value—any formula or procedure or application will use this specific, single value of p at every level for the entire formula or procedure or application.

P-typical Witt vectors don't seem to address how the mind works, or how the brain works, with respect to the mind's and the brain's flexible extent of detailing.

Suppose we're creating categories and sub-categories and sub-sub-categories within a particular field of interest. This is a very p-adic thing to do.

For Digital Mind Math, we do not expect that our p-adically largest category is the parent of a family of exactly p sub-categories, that each in turn have the same number p of sub-sub-categories, as a p-typical mathematical development would suggest. This would be too restrictive a formulation, to require every level of categorization to have the same number of sub-categories.

Universal Witt vectors take p-typical Witt vectors a step further, by permitting formulas or procedures or applications to apply for multiple values of p simultaneously, not just for a single value of p.

And universal Witt vectors (also called "big Witt vectors") can be shown to form a mathematical system that works just like p-typical Witt vectors: The ring of universal Witt vectors is homomorphic to the ring of p-typical Witt vectors.

The Dutch mathematician Michael Hazewinkel has written

extensively about Witt vectors. For him, "the Witt vector construction is a very beautiful one."[17] About the central mathematical operator for the big Witt vector, Hazewinkel writes: "It seems unlikely that there is any object in mathematics richer and/or more beautiful than this one."[18] He writes later: "To me they are so elegant and natural that they simply cry out for deep study."[19]

The Big De Rham-Witt Complex and Crystalline Cohomology

We have shown that a p-adic number is an ordered labeling of elements of a sequence. Reading from right to left, the digits of a p-adic number list, from largest to smallest, an ordered string of components or events.

We have also shown how vast a space p-adic mathematics operates in, meaning that each p-adic number is capable of representing a vast amount of information.

In Part Three of *Digital Mind Math*, we will discuss many examples of how powerful this p-adic numbering system is as a mathematical model of the mind, with each digit of a p-adic number labeling or representing a four-dimensional spacetime event, a spacetime 4-sphere, a mapping of how objects in three spatial dimensions progress through the fourth dimension, time.

In Digital Mind Math, each digit of a p-adic number represents a spacetime sheet or an event, and a p-adic number is a string or sequence of these digits that together form a thought.

This is all of the p-adic mathematics that will be explicitly used in Part Three. But before we get there, let's wrap up the even larger context within which this already large context of p-adic mathematics exists.

We have shown that mathematicians prefer to use the Witt vector framework (with Teichmüller elements), rather than the power series framework with our familiar counting numbers, to continue their exploration of p-adic mathematics. And we have also shown that we do not need to restrict ourselves to p-typical Witt vectors, but instead may broaden ourselves to universal Witt vectors—also known as big Witt vectors—which permit us not to worry about each digit having the exact same number p of possible values.

But this is not the end of the story for mathematicians. Mathematicians look at sheaves of Witt vectors—collections of compatible mathematical information, bundled together like stalks of harvested wheat—in what is called the de Rham-Witt complex. Mathematicians here invoke the image of a crystal—crystalline cohomology—to describe the elaborate connectedness of of this de Rham-Witt complex mathematical construction. The de Rham-Witt complex is a topological tool in the field of crystalline cohomology.[20, 21, 22]

Mathematicians further develop this construction so that it is not limited to a specific prime number p, but rather so that it can be used for multiple prime numbers at once, as a universal or big de Rham-Witt complex.[23]

This mathematics—sheaves of universal Witt vectors—is many stages beyond anything we'll be discussing in *Digital Mind Math*. But I mention it here because it's important to know that it exists. It is important to know that there is a mathematically sound and consistent framework for identifying p-adic numbers in the crystalline context of all other p-adic numbers that they interconnect with—sheaves of p-adic numbers intricately linked with other sheaves.

So in Part Three—as we gain familiarity with how the mind uses p-adic mathematics to self-organize and to generate possible thoughts and select next thoughts—always remember that each thought comes with a crystalline linking of sheaves of interconnected thoughts.

After all, although we're discussing the mathematics of the

mind, we'd like this to be a brain-ready mind, a mind that fits right onto the brain and its billions of neurons with thousands of billions of interconnections.

P-adic mathematics in its full realization is rich, elegant, and natural—the natural mathematics of the mind.

No False Steps

Hensel's Lemma gives us one other element of p-adic mathematics that plays an important role in giving p-adic mathematics the advantages it has as the mathematics of thinking, the mathematics of the mind.

Kurt Hensel was a German mathematician who, late in the nineteenth century, was the original developer of p-adic mathematics.

A lemma is a mathematical theorem. The reason it is called a lemma, rather than a theorem, is as a kind of demotion, in the sense that it is thought not to have particular significance in its own right. Instead, a lemma is a digestible chunk within the proof of a larger theorem that for clarity of argument has been segmented as a subservient theorem, only a lemma.

For purposes of Digital Mind Math, I would like to promote Hensel's Lemma to a full theorem, due to its importance in its own right. To understand why, we need to first digress to an approximation technique called Newton's method, which is frequently used in real mathematics to find numerical solutions to equations.

You probably recall from high school algebra that you were asked to find the solutions (the values of x that make the equation true) to equations such as:

$$x^2 + x - 6 = 0$$

The solutions were obtained by first factoring the equation into:

$$(x-2)(x+3) = 0,$$

then noting that this must mean that either x - 2 = 0, or x + 3 = 0. In other words, there are two solutions to the equation: x = 2 or x = -3.

It turns out that, for computers, this isn't a particularly efficient way to solve equations, and in any case most equations don't factor so simply or have such direct solutions.

So computers don't attempt to solve equations by factoring, and instead generally use approximation methods, such as Newton's method, to obtain precise numerical solutions.

Newton's method starts with an estimated solution, then uses a certain approximation technique to get closer and closer to the solution, usually pretty rapidly converging on the solution.

The only problem with Newton's method is that it doesn't always converge on the solution. In fact, the initially estimated solution needs to be in a vague sweet spot in order for the method to converge on the solution. If the initial estimate is not in this sweet spot, Newton's method won't converge. This is problematic because of how circular it is: If we don't know the solution, how can we know whether an estimate is in a sweet spot close to the solution?

The above description of Newton's method relates to the use of Newton's method for real numbers. What Hensel's Lemma proves is that Newton's method, when applied to p-adic mathematics, always converges. Newton's method, applied to p-adic mathematics, always converges on the solution, and does not require that the original estimate be close to the unknown solution.

I take this as important to the use of p-adic mathematics as mind mathematics—to an evolutionary advantage obtained by a mind/brain complex using p-adic mathematics—because it means that any thought that increases local, momentary knowledge always increases global knowledge. There are no false steps.

So, using p-adic mathematics, we are guaranteed that we are advancing our long-term information content if we advance our short-term information content.

Yes, some local, short-term thinking will advance long-term thinking faster than other thoughts. But we can be comforted in knowing that there are no backward steps, no total wastes of time. This cannot be said for a real-mathematics-based thinking process.

I take this as another reason that the mind has evolved to use p-adic mathematics as the mathematics of thought, a reason beyond the great advantage already established deriving from the vast informational space in which p-adic mathematics operates.

By the way, two paragraphs back, we conceded that some local, short-term thinking will advance long-term thinking faster than other thoughts. We will return to this point in Part Four, where, deep into the intersection of p-adic mathematics and Digital Mind Math, we will speculate on a mathematical formula for ensuring that our short-term thinking maximizes the advancement of long-term thinking—for ensuring that each thought is the best that it can be. This relies on some fine print in exactly what is meant by Hensel's Lemma, for which in p-adic mathematics (unlike in real mathematics) convergence is guaranteed without conditions on the polynomial for which we're trying to find the root, but for which there are nevertheless certain "universal conditions" that must be met. It's too early to discuss this now. So keep reading . . .

The Natural Mathematics of Cognition

Before we move on to Part Three, in which we will apply Digital Mind Math to a variety of ordinary experiences, I conclude with one clarification.

I am in no way suggesting that any of us actually has to understand p-adic mathematics in order to think, in order for our minds to function according to p-adic mathematics. Only a few hundred people in the world understand all of p-adic mathematics.

Here one is reminded of Piaget's explanation for the fact that a child is able to think logically before he has a formal understanding of logic:

> "Formal logic in the current adult sense of the term (I mean the ability to reason logically as Molière's character M. Jourdain did when he triumphantly discovered that he had been talking prose all his life without knowing it, and without the understanding of such a discipline) is not present until ages 11 or 12 through 14 or 15."[24]

Just as Molière's Monsieur Jourdain spoke prose from a very early age without an understanding of the discipline, so too, according to Digital Mind Math, do we think p-adically even without knowing p-adic mathematics. Instead, the wiring of our brains, through trial and error, has evolved p-adically, has evolved consistent with the natural laws of p-adic mathematics.

Part Two's introduction to p-adic mathematics is not a postdoctoral thesis. In fact, listed in the Appendix are a number of theoretical mathematical issues whose resolution and development can help push forward the practical implementation of Digital Mind Math's p-adic mathematics. But because a premise of Digital Mind Math is that p-adic mathematics is our first mathematics and the natural mathematics of cognition, it is useful to ground ourselves in an intuition for the nature of this mathematical system, even if this is a great simplification of a highly advanced branch of mathematics.

Now on to our examples.

PART THREE: ANECDOTES AND ANALYSIS

THE MECHANICS OF DIGITAL MIND MATH

THE ORGANIZATION OF THE MIND
- Grouping details to form concepts
  - The boundary of a thought
    - Complete thoughts
      - Linking concepts
- $M^4_+ \times CP_2$ structure: Spacetime 4-spheres, intricately linked
- P-adic structure: Sequences of numbered 4-spheres and their connections

THE CORE COGNITIVE QUANTUM PROCESS
- Possibility generation: Generation of possible next thoughts
  - Adding detail
  - Migrating to a connecting thought
    - Selection: Selection of the next thought
  - Guided by the maximization of information content
  - Negentropy maximization is the only value
    - Experience: Experiencing the selected thought

IN PART THREE:
We describe how everyday events can be understood through the mechanics of Digital Mind Math.

Each chapter of Part Three will be introduced by identifying which aspects of Digital Mind Math are illustrated by the chapter's anecdotes: the organization of the mind; the core quantum process that Digital Mind Math obeys—possibility generation, selection, experience; or some component or combination of these aspects of Digital Mind Math. Then, at the end of each chapter, we will review what this chapter says about Digital Mind Math.

The human brain has a trillion neurons, which form a quadrillion synaptic connections.[25]

This is 10^{12} neurons. 1,000,000,000,000 neurons.

And 10^{15} connections among these neurons. 1,000,000,000,000,000 synapses (synaptic connections).

It is the premise of this book, *Digital Mind Math*, that these connections form naturally, according to the brain's natural mathematics, p-adic mathematics, the mathematics of cognition.

Our brain accomplishes this without understanding how it is accomplishing this. It is a natural process, evolutionarily developed.

But if we want to understand this—if we human beings want to understand what the brain does naturally as an organ of our bodies—we need to understand Digital Mind Math, p-adic mathematics applied to reconfigured $M^4_+ \times CP_2$ spacetime.

Only p-adic mathematics—in its full realization—will allow us to understand how the brain works. It's our only hope for mapping and understanding a quadrillion connections.

This is because each p-adic number represents a large quantity of information within the vast informational space Ω, and because unlimited quantities of these p-adic numbers and their intricate interconnections may be simultaneously manipulated and analyzed as sheaves of Witt vectors in the big de Rham-Witt complex's crystalline cohomology.

All of this is possible only because of the efficiency of Topological Geometrodynamics' reconfiguration of space and time into a many-sheeted spacetime that is the natural geometry of p-adic mathematics, permitting quantum physics' core process, the quantum jump, to be the natural mechanism of thinking.

It is my claim that—in spite of the many-syllable advanced concepts in the previous two paragraphs—this is all a simple, intuitively appealing process. The purpose of Part Three is to convince you of this.

Personally, because I have spent a decade studying in great depth both the mathematics and the physics that underlie Digital Mind Math, I can attest to the fact that, in my day-to-day life, throughout the day, I observe my mind thinking p-adically.

It is this intuition that I will attempt to convey to you during the course of Part Three.

My goal is that you too see that Digital Mind Math is the natural mathematics of cognition, and that you too will be able, throughout your day, to see this operating as your mind's natural mathematics.

For some of you, the way your mind works is that you are already there, already able to see Digital Mind Math as your own mind's way of thinking, just by Part One's explicit discussion of the mechanics of Digital Mind Math, and Part Two's explicit discussion of the mechanics of p-adic mathematics. If so, Part Three will help confirm your understanding and give you practice. Personally, this has been my way into an intuition for Digital Mind Math: Understanding the technical mechanics and mathematics has allowed me to envision the operational process.

But for many readers, you may not be there yet. Perhaps you learn best from examples. Or perhaps you haven't yet internalized the mathematics and phsycis of Digital Mind Math. If so, let's hope that Part Three's anecdotes and analysis will get you there.

One warning: You will not finish this book with a cookbook recipe for mapping a quadrillion synapses, or for artifical intelligence capable of understanding the universe and all of its particles. This is because there are many layers of intellectual achievement to master before this cookbook recipe is written:

- P-adic mathematics is rarely taught and rarely learned, in spite of its great intuitive appeal once it is understood. It's pretty much unheard of as a school subject through high school and most colleges, with rare exceptions.

- P-adic mathematics is a topic of active graduate and post-graduate study, but it is not fully formulated or fully understood.
- Not everyone understands quantum physics.
- Toplogical Geometrodynamics—the basis of Digital Mind Math—is a radical revision of particle physics, cosmology, biophysics, and consiousness studies, which its developer, Finnish physicist Matti Pitkänen, has documented in thousands of pages of writings.

Said another way: "Rocket Science Meets Brain Surgery @ Digital Mind Math."

The mathematics of Digital Mind Math is at the intersection of two of the twentieth and twenty-first centuries' most challenging, advanced, and important fields of scientific inquiry.

It is my hope that this book will provide incentive for a full, practical implementation of the Digital Mind Math principles.

In the meantime, here's your intuitive path. These are ways to practice p-adic mathematics and practice Digital Mind Math, not to become subject experts in technology or psychology or literature or the performing arts. Have some fun applying p-adic mathematics to these examples from everyday life.

Chapter 13. YouTube and Google Video Search. We show how the organization of online video segments available for viewing selection is remarkably similar to the organization of the mind, as modeled in Digital Mind Math. And the process of selecting and viewing videos resembles Digital Mind Math's core cognitive quantum process.

Chapter 14. Pandora. The music streaming service Pandora does a remarkable job of grouping the details of a song to form a holistic concept. And Pandora generates possible next songs, and selects songs for us to experience, based on the principle of offering moderately novel information content for us to experience. Pandora also offers us the opportunity to affirm or contradict its selection, based on how the selection fits into the context of our own personal history of cognitive quantum jumps.

Chapter 15. Magicians. What fascinates us about watching a magician fool us? We look at our drive to figure out the magician's trickery—and even our drive to be tricked—as examples of the natural tendency toward information maximization.

Chapter 16. Boundaries. We examinine how easily we create boundaries for concepts, and how adept we are at working with the interconnections among bounded concepts.

Chapter 17.The High-P Personal Style vs. the Urge to Adjust the Informational Framework. We look at personality traits and styles of interpersonal interaction that exhibit characteristics of high-p or high-negative-n p-adic tendencies, as well as traits and styles associated with the tendency to modify one's informational framework.

Chapter 18. Portals. The physical representation of Topological Geometrodynamics' CP_2 four-dimensional space is illustrated by a video game and by a military strategy.

Chapter 19. Dogs and Phner: Living the Four-Dimensional Life. The life of a dog and the life of a science fiction character show what it means for each moment to be experienced four-dimensionally, as a Topological Geometrodynamics M^4_+ moment.

Chapter 20. Déjà Vu. We look at phenomena such as the déjà vu as Digital Mind Math processes, and explore functional and dysfunctinal aspects.

Chapter 21. Mrs. Wyatt Earp. Reacting to an incidental newspaper reference illustrates the Digital Mind Math core cognitive process and the continual evolution of the organization of the mind.

Chapter 22. The Obsessive-Compulsive Impulse. We look at the urge to complete a thought, and at the advantages that this urge creates.

Chapter 23. The Absent-Minded Professor. This at times comical figure exhibits high-p characteristics: master of details within his area of expertise, but at times inept at making connections, especially in areas outside of his specialty.

Chapter 24. Jokes I—Finding the Boundary. A silly joke illustrates the mind's urge to satisfactorily complete thoughts by grouping details within a bounded concept.

Chapter 25. Jokes II—To a P-Adically Larger Bounded Space. A second joke illustrates how our mind migrates to a component of the most recent thought, a p-adically larger informational space.

Chapter 26.Mysteries I—Finding the Boundary. Some mystery stories reward us for identifying the boundary around the clues, completing the story around the elements of the plot.

Chapter 27. Mysteries II—To a P-Adically Larger Bounded Space. The appeal of other movies is through a plot twist that brings the story to a p-adically larger bounded space by adding in a late detail.

Chapter 28. Where Is I? The core cognitive quantum process culminates in the experience by our conscious self of the selected thought. Each of us—where "I" is—is the accumulation of these conscious moments.

Chapter 29. Informational Structure and Its Weightings. The core cognitive quantum process adopts quantum physics' model of an experienced path selected from all possible paths, which have been probabilistically weighted based on our entire past history.

Chapter 30. Thinking, Considering, Deciding, Planning, Organizing, Wishing, Regretting, Conversing. Everyday cognitive experiences illustrate various aspects of Digital Mind Math's organizational structure and cognitive processes.

Chapter 31. Creativity and Innovation. We look at Digital Mind Math's process of possibility generation, and see how it can at times lead to disruptive novelty and innovation.

Chapter 32. Shared P-Adic Structures: Religions, Philosophies, Memes,Languages. Cultural phenomena are structured like Digital Mind Math's concepts. Individuals incorporate these structures within their own minds in order to participate in shared cultural memes.

Chapter 33. How to Solve Personal and Interpersonal Problems. Maximization of information content is the only value. Therefore, personal and interpersonal problems are solved by increasing information—the information inside individuals' minds, and the information that individuals share.

13. YouTube and Google Video Search

THE MECHANICS OF DIGITAL MIND MATH

THE ORGANIZATION OF THE MIND
- Grouping details to form concepts
  - Linking concepts

- $M^4_+ \times CP_2$ structure: Spacetime 4-spheres, intricately linked
- P-adic structure: Sequences of numbered 4-spheres and their connections
---

THE CORE COGNITIVE QUANTUM PROCESS

- **Possibility generation: Generation of possible next thoughts**
  - **Selection: Selection of the next thought**
  - **Experience: Experiencing the selected thought**

IN THIS CHAPTER:
This chapter looks at how video segments, available through popular video search and video access sites, are labeled and linked. The video segments display M^4_+ events, and the linking structure (CP_2 structure) is most richly modeled p-adically.

We also look at the user process involved in video search sites, showing how it reflects the core cognitive quantum process.

We'll kick off Part Three's anecdotes and analysis by looking at the popular YouTube and Google Video Search applications, examining their striking similarities to Digital Mind Math's structure of the mind. We'll identify what's similar and what's different. And we'll look not only at the similarities and differences with respect to the organization of the mind, but also with respect to how the YouTube or Google Video Search user experience compares to the process of thinking as structured according to Digital Mind Math.

The Structure of YouTube, Google Video Search, and the Mind

Both the YouTube and Google Video Search websites are structured with multiple categorizations. The user can select categories, such as sports, or movies, or most popular. The user can accept the websites' recommendations or proceed according to a personal preference. The user can search on keywords that relate to how the videos are tagged. The website algorithms also produce search results based on statistical correlations that go beyond simply matching search words to video tags, creating recommendations based on how you and others have proceeded in similar circumstances.

This YouTube and Google Video Search structure is remarkably similar to Digital Mind Math's structure of the mind.

In Topological Geometrodynamics, and therefore in Digital Mind Math, the $M^4_+ \times CP_2$ structure of spacetime represents intricately interconnected moving images: Four-dimensional spacetime segments M^4_+ represent three-dimensional objects as they proceed over the course of a segment of the fourth dimension, time. And these M^4_+ spacetime 4-spheres are connected with every other M^4_+ spacetime 4-sphere that they relate to, linked by wormholes in four-dimensional CP_2 space, a space which permits unlimited linking.

The structure of the video websites' video clips, accessible through their various search and selection options, can be thought of as Digital Mind Math in action.

The websites provide choices for how users may access videos—by category, by recommendation, by search. So the access process encloses these choices, which each in turn encloses further options for how the user will proceed.

One user approach is to browse categories, then select a category, then examine the description of videos in that category.

Another user approach is to accept recommendations and make a selection from there.

Another user approach is to move to the search box and enter keywords of interest.

Ultimately, these are all ways to reach an M^4_+ video segment at the end of a category path, or a recommendations path, or a keyword search path.

A particular M^4_+ video segment is the endpoint of multiple approach paths. It is enclosed within the category approach, within the particular category (sports, movies, and so on), as one of a number of examples that can be selected. Or enclosed within the recommendations approach as one of the examples that can be selected. Or enclosed within the structure of the search approach, reached by CP_2 wormhole connections that best lead to a particular M^4_+ video clip.

This $M^4_+ \times CP_2$ structure is a real structure that will ultimately lead us to the real experience of viewing the selected video segment. The software underlying the website must have real links that locate this video and ultimately permit a real macrotemporal display of macrospatial objects in action.

And this $M^4_+ \times CP_2$ structure is also a p-adic structure, which in Digital Mind Math we have personally labeled p-adically for our use. The website's underlying software has its own organizational labeling. Digital Mind Math puts forth the claim that organizational labeling of massive amounts of data is best accomplished using the great efficiency of p-adic mathematics, the mathematics of cognition, in which p-adic numbers operate within a vast, interconnected, labeled informational space.

For example, according to Digital Mind Math, we may proceed along the p-adically labeled path 0.1 YouTube, rather than the p-adically labeled path 0.2 Google Video Search.

Then we may proceed along the YouTube categories approach 0.11, rather than the YouTube recommendations approach 0.12, or the YouTube keyword search approach 0.13.

We select the popular category 0.111, rather than sports 0.112 or music 0.113.

And it's the seventh popular video 0.1117 that strikes our fancy, so that's the one we watch.

Now we've changed the structure of the YouTube mapping, just as a thought changes the structure of our mind's mapping: YouTube updates the view count of our selected video, reflecting our watch time and our user engagement. And YouTube updates its personal record of what we viewed, and the statistics for the path according to which the selected video was reached.

This proces is analogous to how our mind's history of selections that we experience creates new links and strengthens links that had already been present.

Incrementally, YouTube's massive interconnections to this video have now strengthened along paths that correlate with our own profile, since these paths represent the character traits and past history of this one additional viewer of this particular video.

Google search. Now instead of p-adic path 0.1, going directly to YouTube, we could have elected p-adic path 0.2: Use Google's search feature to find videos that interest us. This might lead us to YouTube videos or to videos on other sites.

The Google search process also bears a remarkable resemblance to the Digital Mind Math mapping of the mind. For one thing, there is the real web itself, which we will ultimately access, just as the Digital Mind Math cycle continually brings us to the the real experience of a selected thought. But, similar to the Digital Mind Math cycle's steps of possibility generation and selection, which proceed using p-adic mathematics not real mathematics, the Google search process does not search the web itself—it searches Google's index of the web.

So right away, we again see Google search's dually labeled structure, in analogy to the adelic mathematical structure—both real and p-adic—of Digital Mind Math. Underlying the Google search is a web filled with articles and images and, in this case, videos. But Google has separately indexed the web, intricately mapping the interconnections of videos and what they're about and how they're tagged and who views them and what else these viewers view.

The web is out there, in all of its real glory, ready to be experienced. Google has indexed the web, mapping connections and mapping which videos enclose which content.

The structure of video search sites such as YouTube and Google Video Search is remarkably similar to the Digital Mind Math structure of the mind:

- An index or mapping labels the content that the videos enclose . . .
- . . . and also interconnects every video with every other video that shares themes or subjects or important words.
- The index or mapping is not the web itself; it is an indexed, mapped set of interconnected labels of material on the web.
- When desired, the user may click to view the real video, as a macrotemporal and macrospatial experience whose label we have been led to.

The Process of Thinking

In Digital Mind Math, thinking proceeds according to quantum physics' core elementary particle process, the quantum jump:

(1) Generation of possible next thoughts, by:
    (a) Adding detail, or
    (b) Migrating to a connecting thought.
(2) Selection of the next thought, guided by the maximization of information content.
(3) Experiencing the selected thought.

When we use Google Video Search, we enhance our mind's natural Digital Mind Math capabilities by taking advantage of what Google's elaborate search engine can do for us.

Google enhances our generation of possibilities for the next thought by generating possible videos for us to view. It does this by giving us a very human-friendly user interface: We can just type in words that describe what we're interested in. Then Google gets to work, selecting hundreds or thousands of possibilities that incorporate our keyword concepts, that use or relate in some way to the search words that we typed in.

And Google also helps us in our process of selecting a video, based on the maximization of information content.

In the case of a Google web search for a relevant prose article or document, Google selects articles based on multivariate algorithms that look for our search terms, counts how many times they appear in each article, rates the article higher (that is, ascribes a higher information content) if our search words are in the title, or if the website is of high quality or rank, or if it's referenced frequently.

Video search is similar, although video search must inherently relie less on content words, and more on tags and observed interconnections.

And we, as the user, have work to do too; Google doesn't do all the work for us. Google gives us its ranking of possible videos to view, based on its algorithmic analysis of information maximization. But Google recognizes that it can only go so far with its algorithm: The personal, human element is needed to make the actual final selection.

We then experience the selected video. We watch and listen to the real video, not just a description of the video drawn from Google's index of the web.

Again, the Google Video Search process, when considered in combination with the human efforts, is a process that describes remarkably well Digital Mind Math's model of how we think.

If we searched for a video using YouTube's search feature, the process is very similar to the Google Video Search process of sharing responsibilities with the human user. Depending on the particular YouTube approach that we use, there may be a somewhat different split of responsibilities between what YouTube does and what the human user does.

For example, if we proceed along YouTube's categories approach, we (the human user) will do more work, since the videos within each category will not be presented in an order corresponding to the specific interest we have today, because YouTube's ordering within categories is reflecting general user interests, rather than our own specific interests.

What This Chapter Says About Digital Mind Math

The mechanics of web-based video sites are a lot like the mechanics of Digital Mind Math, both in terms of the structure of how the videos and their interconnections are labeled and mapped, and also in terms of the process according to which we proceed to select and experience the videos.

Each video is labeled with a set of descriptive tags. Each of these tags also appears as the tag for other videos, identifying aspects of similarities among videos in a way that is analogous to the linking structure of Digital Mind Math.

And the video sites' user experience—generating possibilities for the the next video to watch, then selecting and watching the video—resembles Digital Mind Math's core cognitive quantum process.

There is a dual labeling and storage process for web-based video sites: Possibilities are generated and analyzed using video tags and the pattern of tags. But upon selection, it is a real video stream, not just its label, that is experienced. This mirrors Digital Mind Math's use of p-adic mathematics to label and analyze our mind's sequences of concepts (spacetime 4-spheres) and their connections, leading to the experience of the selected next thought in real four-dimensional spacetime.

In the next chapter, we will see how the user experience associated with another everyday phenomenon exhibits characteristics of Digital Mind Math's core quantum process. But this time, instead of focusing on the overall dual p-adic and real informational structure, we will focus more specifically on how details are grouped to form concepts.

14. Pandora

THE MECHANICS OF DIGITAL MIND MATH

THE ORGANIZATION OF THE MIND

• **Grouping details to form concepts**

• Linking concepts
• $M^4_+ \times CP_2$ structure
• P-adic structure

THE CORE COGNITIVE QUANTUM PROCESS

• **Possibility generation**   • **Selection**   o **Guided by the maximization of information content**   o **Negentropy maximization is the only value**   • **Experience**   **IN THIS CHAPTER:**   Pandora is music for the mind. This music service has access to a vast library of songs which it has anaylzed according to hundreds of criteria. Pandora selects songs for its listeners by offering moderate novelty for the listener to experience. Pandora's "music genomes" illustrate the creation of conceptual wholes from groupings of details. And the Pandora user experience is similar to the mind's process of cognition.

Pandora® is a music streaming service which offers listeners the opportunity to hear songs they know and like, as well as songs they don't yet know but are predicted to like. Pandora's predictions have a lot of similarities to what a particular listener has heard in the past and has rated favorably, but these selections also stretch the listener a bit, leading the listener to new artists and songs, and perhaps bringing the listener to new spheres of interest and music appreciation.

The Music Genome Project

Pandora has developed a musical analysis and classification process—a "taxonomy of musical information"—that it calls the Music Genome Project®.[26] Pandora describes this as the most sophisticated taxonomy of musical information ever collected, rating each musical work on up to 450 musical characteristics, such as melody, harmony, instrumentation, rhythm, vocals, and lyrics.

The result is a remarkably resonant descriptive labeling of each song, each musical work. Pandora's labelings—its groupings of details—identify the essence of the song. The labels describe the song, even though they are not the song itself. They are labels, not music.

There is much to note here as similarities between the mechanics of Pandora's Music Genome Project and the mechanics of Digital Mind Math.

Based on the listener's past history of songs that she has listened to, Pandora generates possible next songs, not precisely identical to the previous song, but varying in some details, or connecting the listener in a new direction.

To maximize listener enjoyment, Pandora stretches the listener—moderately, just a little, not a lot. Maybe similar lyrics, but somewhat different instrumentation. Maybe familiar rhythm, but different melody.

The listener gets to tell Pandora if Pandora got it right. The listener can like with a thumbs up, or dislike with a thumbs down. Or the listener can refrain from opining and see what Pandora comes up with next.

Pandora maximizes the listener's musical enjoyment by giving the listener what she wants, with lots of grounding in the familiar, but also with some moderate stretching of the listener's experience in directions that the listener can help shape.

The Role of Human Analysis

Pandora is quite clear that songs are analyzed by trained music analysts, not by automated listening processes. We may ask ourselves: Why have machines equaled or exceeded human abilities for fingerprint analysis, or for facial recognition, but not for music appreciation?

It may seem like ancient history to anyone born into the computer age, but the reality is that it was not too long ago that trained human intervention was required for fingerprint analysis. Today we can enter the country by placing our fingertips on a screen so that a machine can compare our fingerprints to our stored Global Access records.

What is it about fingerprint analysis that permits its automation?

And who can doubt that, within a few years if not already today, closed circuit television capabilities will permit instant identification of the thief who just smashed the window of your parked car and stole your Christmas purchases?

Before increased computer capabilities proved us wrong, we were sure that only trained experts could correctly analyze fingerprints, and that only the unique capabilities of our minds permit us to recognize human faces.

Is the difference that only 20 or 30 measurements uniquely identify a fingerprint of a human face, but it takes 450 measurements to uniquely identify a song?

If so, then, once we have several more leaps in computer

processing capability, Pandora should be able to rely on machine analysis rather than trained human labeling.

But don't forget that Pandora's human intervention is not only at the Pandora end of the process, it's also at the listener's end. Once Pandora selects a new song for us to listen to, we get to veto it if Pandora hasn't given us a song we like.

Fingerprints uniquely identify their owner. And so does facial appearance. But is appreciation of particular songs the same thing? Is an individual's sense of music appreciation an objective characteristic of the individual, as objective as her fingerprints or her facial characteristcs?

The Digital Mind Math focus here is not so much on the human role at the Pandora end analyzing the music, but rather at the listener's end with a thumbs up or thumbs down evaluating the music she was presented.

As subtle and complex as are questions of music appreciation, it seems difficult to argue that machines will forever be incapable of scoring songs along even 450 musical dimensions. True, this is beyond today's computer capabilities. But can we not envision computer recognition of lyrics? Of vocals? Of instruments? Of patterns of melody and harmony and rhythm?

To argue against this seems entirely reminiscent of earlier arguments against computers being able to recognize fingerprints or faces.

But what is quite different is the role of the listener's sense of appreciation, what the listener likes and doesn't like.

Pandora offers us a new song, with lots of similarities to what we've listened to in the past, but with some slight tweaks along a handful of dimensions.

Pandora needs the listener to evaluate whether Pandora correctly anticipated the listener's appreciation for the new musical characteristics it selected.

Pandora needs us to tell it if we enjoyed the selection, if we appreciated the novelty.

Here we see the importance that Digital Mind Math gives to the history of the individual's prior cognitive quantum jumps.

Our prior cognitive quantum jumps have formed connections that group details into concepts, and that form and stregthen links among concepts.

The Digital Mind Math perspective is that we select the next thought to experience—select the next song to listen to—guided by the maximization of information content.

Pandora's expertise does a lot of the work for us. Pandora is expert at generating possible next songs by examining our past songs and adding a detail here or there, moving us to a moderately novel direction or new connection. In so doing, Pandora is performing its own calculation of how we might best maximize our enjoyment by offering us its evaluation of the ideal next song to experience.

Perhaps, guided by our entire lifetime history of cognitive quantum jumps, which have formed our mind's connections, we accept the song that Pandora has selected for us and experience that song, in its entirety, for its entire length, with all our heart, with all our listening intensity, to appreciate the song and everything it has to offer us, to strenthen existing cognitive connections, to create new cognitive connections, to add detail to concepts that are already present within the cognitive mapping that our entire history of cognitve quantum jumps has created for us.

Maybe the song has been so great for us—has so maximized our information content, in whatever way our prior cognitive history has defined this—that we tell Pandora we like this song, thumbs up, give us more, we want to further strengthen this cognitive connection.

Or maybe—in ways only we can know, in ways that Pandora could not have anticipated—Pandora's conjecture did not resonate with us. The moderate novelty selected by Pandora was not novelty in the direction of our desires. Thumbs down. We don't like this. Try a different direction.

Pandora has brilliantly grouped details and captured the essence of our musical history in order to create a concept of our musical preferences. And Pandora has exceeded any capabilities we could possibly have in generating possibilities for our next listening experience. But Pandora needs us to shape its understanding of how we—personally, individually, based on our entire past history of experiences, likes, and dislikes—evaluate enjoyment, evaluate, in Digital Mind Math terms, our extent of information maximization.

Pandora has actually completed for us the entire three-step core cognitive process. It has generated possibilities, better than we could do on our own. It has selected our next thought (our next song) and given it to us to experience. But Pandora can only guess at the selection, and we can provide feedback to Pandora only by experiencing Pandora's guess.

So, as of the early twenty-first century, the combined Pandora/human complex has provided the full cognitive quantum process, but with mixed results: Possibility generation, computer-aided, is exceptional and far exceeds what we could accomplish by human means alone. But Pandora's selection, while a very good guess, may be wrong, maybe very wrong, wrong enough for a thumbs down. And we will be able to evaluate the selection, and provide feedback on the selection, only by experiencing the selected thought (listening to the Pandora-selected song).

Music appreciation is a complex and subtle experience. The more intimately today's computer-aided human capabilities can identify the individual listener's precise history of prior cognitive quantum jumps, the more accurate will be the experience of information-maximizing musical pleasure.

What This Chapter Says About Digital Mind Math

Pandora labels songs according to hundreds of musical criteria. This is analogous to Digital Mind Math's model of the organization of the mind, in which details are grouped to form concepts. Each criterion—each p-adic digit, in Digital Mind Math—is a label of a real experience. Pandora's criteria are grouped to form the musical conception of each song.

It is Pandora, rather than the subscriber herself, that generates possibilities for the next song, then selects and plays this song. In Digital Mind Math, the selection is guided by the maximization of information content. Pandora uses moderate novelty as a proxy for information maximization. We have seen that Piaget also noted the attraction toward moderate novelty in children's path toward higher and higher levels of equilibration. Vygotsky labels this— Pandora's next song—as lying in the zone of proximal development.

The Lovelace test. Will Knight, senior editor for artificial intelligence at *MIT Technology Review*, recently reviewed the state of the music streaming industry, with special note of Apple Music's hiring of hundreds of DJs to create playlists, in supplement to algorithmic techniques. We have already mentioned Pandora's active use of human intervention to rate songs according to the Music Genome Project's musical criteria, but Knight suggests the possibility that enhancing human intervention at the front end—not just rating songs, but also selecting songs—offers the possibility of enhancing musical selections' "emotional resonance and cultural context" with a level of creativity beyond what algorithmic recommendations can offer.[27]

Knight discusses other approaches to algorithmic musical recommendation, such as Spotify's "collaborative filtering" methodolgy, which compares a user's listening behavior to data collected from other users, and Spotify's deep-learning network, which goes beyond songs' metadata to directly recognize audio features. But ultimately, Knight feels that success is to be measured by passing an auditory Turing test, a creativity test—similar to the Lovelace test, developed by two Rensselaer Polytechnic Institute researchers along with IBM's lead Watson developer, David Ferrucci. This test is named after Ada Lovelace, the pioneering nineteenth-century mathematician instrumental in the earliest formulations of modern computer science, who expressed skepticism that computers could ever do anything original. A musical Lovelace test—similar to the Turing test of whether a user can distinguish a conversation with a person from a conversation with a computer— will see if listeners can distinguish a computer-generated playlist from one developed by a human DJ.[28]

Don't lose sight of the fact that Pandora's Music Genome Project was introduced at this point in *Digital Mind Math* because of how well it illustrates aspects of the Digital Mind Math model of how the mind is organized and how the core cognitive quantum process proceeds. However, this discussion has also raised questions about how creative any algorithmic process can be, and therefore this allows us to segue to the topic of the centrality of negentropy maximization (maximization of information content) in Digital Mind Math.

Information maximization. You may not yet be convinced that the pleasure of listening to just the right song can be reduced to information maximization, or, in the terminology of TGD physics, to negentropy maximization, to the greatest negative entropy, to the greatest quantification of order and organization.

It may take many more chapters for you to feel comfortable with Digital Mind Math's only value being negentropy maximization.

We'll continue this in the next chapter, where we will look at a specific example of the urge to maximize information content.

15. Magicians

THE MECHANICS OF DIGITAL MIND MATH

THE ORGANIZATION OF THE MIND
- Grouping details to form concepts
  - Linking concepts
  - $M^4_+ \times CP_2$ structure
  - P-adic structure

THE CORE COGNITIVE QUANTUM PROCESS
- Possibility generation
- Selection: Selection of the next thought
  - **Guided by the maximization of information content**
  - **Negentropy maximization is the only value**

IN THIS CHAPTER:
Our lifeforce drives us to increase our information content. This process is guided by properties of the physical world, properties of physics: maximizing order, maximizing negative entropy. Our brain's structure of neurons and synapses implements this information-maximizing process.

- Experience

Many of us enjoy watching a talented magician perform.

Right before our eyes, with our full knowledge that we're going to be fooled, we scrutinize the magician's every action.

The magician, meanwhile, is deliberately drawing our attention toward something irrelevant, so that she can perform the act of magic without us seeing.

It is thrilling to be fooled by the magician. It is fun.

Our jaws drop in awe of how she pulled this off.

Between acts, and after the show, we try to figure out how the magic trick was accomplished: It was in her other hand. Or between her fingers. Or folded up. Or there was a trap door.

Why do we enjoy watching a magician?

The Aestheticization of "How"

Jenny Hendrix's review of Steven Galloway's 2014 novel *The Confabulist* gives Hendrix the opportunity to discuss the magician's method: An effect, hidden by the magician's method, triggers the audience to attempt a reconstruction afterward. The audience's enjoyment derives from an appreciation of the magician's skillful misdirection—"the aestheticization of 'how.'" [29]

This is information maximization in action. And it is a thrill, a pleasure. We enjoy being fooled, because to be fooled means that we are in the presence of a larger informational space that we have not yet entered. But we're close to. We've been made aware that it's there. And for just the price of admission we have the opportunity to expand our knowledge, increase our information, decrease our entropy, reach a higher level of equilibration.

And there's something playful about this—reminiscent of Vygotsky's imaginative play as our way to actively construct knowledge. We're presented a game, a puzzle. And as a result, we have the opportunity to proceed with some mind work, and ultimately reach closure, at a higher level of equilibration, once we understand how the pieces of the magician's act fit together.

This concept of reaching closure—the concept of the boundaries of a thought—is one that is central to Digital Mind Math: What does it do for us to identify the edges of a thought, the thought's boundaries?

In the next chapter, we will gain more experience in how central boundaries are to how we think, and how good we are about thinking in boundaries.

Then in following chapters, we'll explore more about how the boundaries of a thought allow us to move on to p-adically larger thoughts—that is, how boundaries help us to maximize information.

And we will also explore how the boundaries of a thought make our mind's search process more efficient, how resolved boundaries permit a thought to be a point within a p-adically larger thought, so that we're that much further ahead in the efficiency of our cyclical generation of the configuration space of next possible thoughts.

What This Chapter Says About Digital Mind Math

Underlying the entire Digital Mind Math process is a drive, a life force, to maximize information content, to increase negative entropy. This is the negentropy maximization principle of Topological Geometrodynamics, the physics that underlies Digital Mind Math.

16. Boundaries

THE MECHANICS OF DIGITAL MIND MATH

THE ORGANIZATION OF THE MIND
- Grouping details to form concepts

---

o **The boundary of a thought**
o **Complete thoughts**

IN THIS CHAPTER:
An important element of our understanding a concept is to scope out what the concept includes, and what the concept does not include. When we understand the boundaries of a concept, we are able to reduce the concept to a single point that we can juggle with other concepts and hypothesize how multiple bounded concepts interrelate.

- Linking concepts
- $M^4_+ \times CP_2$ structure
- P-adic structure

THE CORE COGNITIVE QUANTUM PROCESS
- Possibility generation
- Selection
- Experience

In this chapter, we examine how boundaries are an important aspect of how we think: We construct boundaries. We examine the edges of concepts, evaluating the certainty or uncertainty of their boundaries. We conceptually manipulate mentally constructed bounded universes.

These processes are all straightforwardly modeled p-adically, through our Digital Mind Math.

Let's see how.

Parking at the Post Office

I live in a small town, with a centrally located post office, which has a very convenient parking lot located behind it.

It is generally understood that the parking lot is for use only while conducting post office business. This understanding is supported by the signs reading:

15 **MINUTE** **POSTAL** **PATRONS** **ONLY**	and	**ILLEGALLY** **PARKED** **VEHICLES** **WILL BE TOWED** **AT VEHICLE** **OWNER'S** **EXPENSE**

Human nature being what it is, and parking conditions often being quite difficult, and ecological justifications being easy to rationalize (not starting the engine multiple times to move the car for different nearby errands), there is a bit of an unwritten tolerance for parking at the post office and doing a quick additional errand or two.

Most people are not flagrant violators—just minor violators—and they don't want to be perceived as flagrant violators. Even as they also don't want to move the car more than once—even if they could find another parking space—for multiple nearby errands.

So it's natural, when one is parking at the post office with the intention of also doing one or two other quick errands, to consider when one is exposed to adverse judgment from others about parking in the post office lot while doing other errands: When will an observer be able to see the evidence that I am a postal parking lot violator?

The post office is on a one-way street. To reach the parking lot driving along the one-way street, one first drives past the parking lot exit, which is on the left, then drives past the post office itself (which is also on the left), then takes a left turn into the parking lot entrance in order to park in the back behind the post office. Eventually, one will drive out via the parking lot exit, which takes you back out to the one-way street, where one again passes the post office and the parking lot entrance on the left before proceeding on one's way. The pedestrian entrance to the post office is in the front, and pedestrians may walk into or out of the parking lot via either the vehicle entrance or the vehicle exit.

If one is going to the bank next store, before going to the post office then returning to one's car parked behind the post office, one is exposed to a passerby's knowing about the violation of the requirement that parking is only while conducting post office business.

When, exactly, does this exposure begin and end?

Most obviously, when one gets out of the car after parking behind the post office, and walks along the path of the parking lot vehicle exit toward the street and toward the pedestrian entrance to the post office, the instant that one turns right, toward the bank, rather than continuing left, toward the pedestrian entrance to the post office, it becomes plain for all to see that a parking pecadillo has been committed, a violation is in progress.

We know instantly when that moment occurs. We instantly use the boundary to create two conceptual spaces—on one side, before the moment of transition, is the region where no violation is observable; on the other side is the region where the violation is active.

A bounded concept is created in our mind. This concept

represents the region in which we may be seen to be a scofflaw.

When we're seen walking out of the parking lot into the violating region toward the bank, we know that someone could see us in all our legal indifference, and we mark this concept in our mind as a bounded region in which we're exposed to being uncovered as a violator.

Note in particular that the boundaries of the concept correspond quite precisely to measurable boundaries on the ground. We could identify the concept as 4 feet, 7 inches from the edge of the street, 16 feet, 11 inches, from the entrance to the post office. But we don't. The identification is conceptual, not metric. We fix the boundary of the concept in our mind, label the concept by its conceptual edges, not its metric edges.

So we've made two points about Digital Mind Math so far: First, we are very adept at creating, and highly motivated to create, mental concepts that are conceptually bounded spaces, for which we have some kind of internal labeling procedure to name and to be able to reference in the future. This concept, representing the region of exposure, is linked to other concepts, such as the route to the bank, and the risk of being discovered, and the consequences of being discovered, and the options for minimizing risk, and the two-element and three-element and four-element (etc.) combinations with other possible bounded concepts that together could expose us as the violator that we are.

We focus on a concept—the region of exposure—and in experiencing this concept, thinking this thought, the thought brings with it a whole host of connected thoughts.

This model of how we think is the Digital Mind Math model: a master thought, which we experience, and which is related to other thoughts by a p-adically recorded scheme of interconnection.

The second Digital Mind Math point to note is that the concept has an ultrametric label: It is not measured; it is labeled. Said another way, our mind records the concept p-adically, as an ultrametric, labeled description, not by its real (feet and inches) measurements.

Who might catch us? Which passerby might be able to see us in violation?

Most obviously, a passerby who saw us get out of the car, and continued to watch us walk to the front of the parking lot and make the damning right turn toward the bank, will certainly know of our violation.

Again, this creates a bounded concept for us, a concept we label as the region in which an observer could see what we've done. Again, this is a bounded concept, which we could identify and remember by its real physical measurements, but we don't. Instead, we label the region conceptually, and locate it with wormhole connections to other concepts—exposure, guilt, hiding, proceeding briskly, but casually, so as not to draw attention.

We get to the bank, and our friend Denise teases: "I saw you parking at the post office."

How did she know?

We are instantly capable of imagining possible sequences of events that have permitted Denise to know we parked at the post office and went to the bank.

Not by measurements. By concepts. By labeled bounded concepts, that we instantly conceptualize. By a configuration space of possible events that could be the explanation for how Denise knows.

Most simply, maybe she saw us make the fateful right turn, walking out of the post office parking lot, toward the bank, not toward the post office entrance.

But I didn't see her see me . . .

Maybe she sees my stamped unmailed letters in my hand.

But I could just be walking (not driving) to the bank and post office . . .

Maybe she was leaving the card store across the street while I drove by the post office before entering the parking lot.

She's carrying something in a bag from the card store!

"Wish your mother 'Happy Birthday!' for me," is my retort.

This emerges from my probabalistic analysis of the configuration space of possibilities that my mind has generated, possibilities that are logical but that are also based on my prior history of quantum jumps, as they have influenced the neuronal connections in my brain.

I remembered that Denise's mother's birthday is coming up, and what's likely is that it was because she was getting a birthday card across the street that she saw I was in a vehicle driving in front of the post office.

Now it's Denise's turn to generate some combinations of bounded concepts that will explain why I said this.

But let's make this harder, more complex. After all, Denise is a friend, and it isn't a particularly emotion-ridden event for me to be teased by Denise.

But Dennis is not a friend. I had recently told Dennis to stop parking in my driveway. And when I leave the bank and enter the post office, there's Dennis: "Oh . . . Using the post office parking lot to go to the bank too?"

Now my initial thought—the unavoidable wormhole connection, the irresistible segue from Dennis's comment—is that Dennis is making a direct reference to our earlier driveway tiff.

This is a segue controlled by emotion, in which I'm tempted to achieve some kind of greater resolution in my life and its issues by going down the road of furthering the interaction on the issue of parking misdemeanors.

But I resist going down that road. Save it for another day. Or some private thinking.

I'd like to figure out how Dennis knew I was parked in the back of the post office but had gone to the bank first.

This is a bit more complex than the situation with Denise. I am, after all, in the post office, not the bank. And I have mail in my hands, and my car is parked in the the post office parking lot.

The straightforward appearance is totally upright and law-abiding.

Instantly, my mind goes to work, thinking of the possible combination of events that has permitted Dennis's snide comment.

Our minds are masterful at this: In the blink of an eye, we can construct multiple sequences of events that could have exposed our guilt to Dennis. We can instantly conceptualize one-part and two-part and three-part and four-part combinations of activities—sequences of bounded spaces—that would have permitted Dennis to know: simple sequences, and more elaborate sequences.

Was Dennis in the bank? Or did he see me leaving the bank? But even if so, when did he see the car? Is he just assuming I drove? Did he see me drive by the post office? Into the lot? Did he see me walk to the right toward the bank?

Remember: There's a whole bounded region, a whole bounded concept, where it's clear that I am outside of the straight-line path that runs from my car behind the post office directly to the post office pedestrian entrance. Now that Dennis sees me inside the post office, all it would take—in an Occam's Razor search for the simplest solution, which is now a multi-step solution, since a single-step solution is ruled out by the fact that the current appearance of me being inside the post office seems totally law-abiding—would be Dennis having seen me in the region outside of the straight-line path, in combination with something indicating I was parked in the back.

Maybe he saw me leaving my driveway a few minutes ago. Or he saw my car in the back. But did he also see me on the way to the bank, or inside the bank?

Maybe he didn't see me in the bounded region around the bank, but he saw the bank deposit receipt.

There is some sequence of bounded conceptual events that has let Dennis know that I parked in the post office parking lot and proceeded with business other than post office business. That much I know.

Maybe for now I'll file this concept away. It may become useful at some later date to know exactly how Dennis knew.

Until then, I'll just ignore the whole thing. In this circumstance, that's what brings me to the highest informational state.

Purely Conceptual Boundaries

The bounded concepts discussed so far largely correspond to bounded metric areas. But this is not always the case. We routinely also construct conceptual boundaries, boundaries that are not so directly and obviously linked to real metricization.

For example, suppose we wanted to legally justify our parking in the post office parking lot while we proceeded with non-postal errands.

Our first impulse might be to examine the exact wording of the signs: "15 minute postal patrons only" and "Illegally parked vehicles will be towed at vehicle owner's expense."

We have two plain-English signs, that each describe rules and consequences for a specific situation. Each sign describes a concept, and we want to examine the boundaries of these concepts to see what situations will or will not fall within the penalizing bounded areas.

There is no metricization here. This is a boundary creation that is more typical of how we usually define boundaries in order to manipulate concepts—purely conceptually, without real measurements.

Getting around the second sign—"Illegally parked vehicles will be towed at vehicle owner's expense"— is easy: It doesn't define what's illegal. We conclude that the boundary of its punitive relevance is given not by anything in this sign, but rather by the other sign, assuming that the other sign offers a boundary of punitive relevance.

So we conceptually link the operative impact of the second sign to the bounded scope of the first sign: The second sign warns us that we can get towed away—at our own expense—but the circumstances (bounded scope of events or conditions) under which we will get towed are defined by the first sign.

The concept of being towed from the post office parking lot is a refinement of—a p-adically larger condition than—our preexisting car-being-towed concept: If, based on our history of quantum jumps, we have labeled having our car towed as the p-adic event 0.938275, then we need to look at the event 0.9382751, an event that in a real sense is more detailed, which means that in a p-adic sense is p-adically larger.

In our mind, based on our history of cognitive quantum jumps, we have labeled the event of having our car towed as p-adic event number 0.938275. We will now focus on this specific implementation of our car being towed—p-adic event 0.9382751, a p-adically larger event, the event of our car being towed right now from this particular parking lot.

We need to examine event 0.9382751 and its links, its interconnections.

We know that event 0.9382751 is linked to the event of paying a particular fine specified for this circumstance. And we also know that it's linked to the event of creating an inconvenience, making us late for the activities we had planned on doing next. But those aren't the links that we want to examine right now. They won't optimally maximize this moment's information content. Right now, we need to examine the link to the event—the bounded concept—of what constitutes a tow-inducing action at this time in this parking lot. This link will trigger the information-maximizing action of examining the first sign regarding what defines a violation, what events are within the boundaries of violating behavior.

Note that we are now talking about the boundaries of events, but these are only conceptual boundaries, without a clear metricization. Nevertheless, our mind is adept at defining and understanding the implications and interconnections of these events, even though measuring them is vague.

And what does the first sign say? "15 minute postal patrons only."

Here's our opening. Arguably, the boundaries of the concept "15 minute postal patron" seem to be large enough for our purpose. If we limit our total parking time to 15 minutes, and if we are in fact a postal patron, we can put forth the argument that we are complying with the conditions of the sign, even if we also go to the bank.

This will be our argument, our justification, our rationalization, which (maximizing information content) we are preparing just in case there will be an interaction with a tow truck operator: The boundaries of legal parking activity are given by the conditions of (1) parking for a maximum of 15 minutes, along with (2) being a postal patron. Included within these boundaries is the condition of going to the bank, as long as we also patronize the post office and park for a total of no more than 15 minutes.

Interviewing a Job Candidate

Suppose you're interviewing an applicant for a postion in your company.

You have just a brief time to get to know her, so every moment is a moment of evaluation, of putting together a picture of who the candidate is and how she will fit in with your company and add value. We need to be sure, at every moment, that we're maximizing information about this candidate.

The *New York Times* publishes a "Corner Office" column in the SundayBusiness section, in which columnist Adam Bryant interviews business leaders about their leadership style and corporate culture. Bryant asked Ron Kaplan, chief executive of the outdoor deck manufacturer Trex, how Kaplan makes his hiring decisions.

One aspect of Kaplan's interviewing technique is particularly noteworthy: He likes to hand his car keys to the interviewee and ask the interviewee to drive while Kaplan asks business-related questions, in order to see how the interviewee multitasks.

Here's how Kaplan describes why he likes this unusual aspect of the interview:

> "By the time the interview's over, I want to know their character. Do I know this person? Do I have a sense of who they are? If, after 90 minutes with them, they're not well-defined, and I can't see the edges, I get uncomfortable. To be a good leader, you've got to be predictable. And to be predictable, you've got to know the person to a reasonable extent.[30]

Kaplan is looking to locate the edges of his job candidate's personality, the boundaries of the interviewee's leadership style. Kaplan equates this with getting to know the person, with Kaplan developing a sense of predictability about the behavior of this job candidate.

And Kaplan takes this further, to put forth that a successful leader is one that—within the course of a 90-minute period—routinely shows her edges so others can see, demonstrates the boundaries of her style and preferences, makes herself predictable.

This is an example of our penchant for defining boundaries as an important element of how we think, an important element of

Digital Mind Math.

In the next chapter, we will discuss more about the drive to form conceptual boundaries, and about circumstances and personal styles in which concept formation is at the forefront. And we will contrast this drive to group details to form bounded concepts with a competing drive, the drive to link concepts and form conceptual interconconnections.

Boundaries for Psychological Health

Boundaries are a key concept for a number of aspects of psychological health and psychological dysfunction.

Therapist-patient boundaries. Psychotherapists think a lot about therapist-patient boundaries. There are forms of therapy in which the whole point is for the therapist to be a blank slate on which the patient projects her life issues. In such therapy, it is a theoretical precept that a firm boundary be maintained around the patient's cognitive state, and that the therapist's cognitive state be rigidly bounded so as not to interact within the bounded space that the patient constructs.

There are also legal risks associated with a therapist crossing the therapist-patient boundary, by seeing the patient outside of therapy, touching or sexual interactions, accepting gifts, or even disclosing personal information.

On the other hand, some approaches to therapy—for example, some humanistic therapies or cognitive-behavioral therapies—deliberately cross boundaries, for example through the therapist's self-disclosure, as a therapeutic technique.[31]

Boundaries for psychological health. Aside from the boundaries maintained or crossed between the therapist and patient, the patient's learning about boundaries in her own life can be an important issue on the path to psychological health. Try a Google search on "learning about boundaries," and you'll get 127 million hits in 0.36 seconds:

- 10 Ways to Build and Preserve Better Boundaries[32]
- Setting Boundaries with Difficult People[33]
- 6 Steps to Setting Boundaries in Relationships[34]
- How to Set Healthy Boundaries: 3 Crucial First Steps[35]
- 4 Ways to Establish Boundaries[36]
- and 126,999,995 more

Boundaries are created by rules and values and laws. Healthy psychological and interpersonal boundaries keep us emotionally and physically safe. Unhealthy boundaries can keep us from leading a fulfilling life and can derive from a dysfunctional family or childhood background.[37]

Boundaries are a key psychological concept. We routinely form and refine conceptual boundaries, and use boundaries to guide our thinking and actions—and ultimately to maximize information content.

Telescoping, Amplifying, Jumping

Commentators on literature and film note how the construction of a complete bounded concept creates opportunities to make a broader point or to jump to connected points.

For example, the actress Laverne Cox describes how, in her character's portrayal on the Netflix series "Orange Is the New Black," it is the embrace of her specific identity—as a trans black woman from a working-class background in the southern United States—that allows generalization: "When we get specific in the storytelling, that's when the universality happens."[38] It is this grouping of details to form a specific bounded complete concept that, perhaps ironically, allows the link to the general and universal concept.

As another example, consider the description of a literary technique deployed in a collection of short stories that "avoids mawkishness because of its close observation of objects and surroundings . . . [S]tartling images bring a moment to life with searing heat."[39] It is the close observation and narrowing of focus—the creation of a boundary around the details of a concept—that links the concept to somewhere it otherwise could not go, in this case toward fresh and sophisticated themes, away from mawkishness and sentimentality.

And consider this description of James Joyce's literary tecniques in *Ulysses*: "telescoping some moments, amplifying others, jumping from character to character, continent to continent, subject to subject, text analysis to literary history . . . [employing the] technique of 'epiphanies'—profound everyday scenes—to create fresh pictures."[40] Again, it is the profoundly described everyday scene that creates the epiphany, the fresh picture: Precisely because of the profound specific description are we transported to a fresh picture, in spite of the scene being everyday. And this description of Joyce's techniques gives us more Digital Mind Math applications— telescoping, amplifying, jumping—that result from richly described details grouped together to form a complete bounded thought: In Digital Mind Math, we generate possibilities for our next thought by telescoping to small details, or amplifying a concept that we expand upon, or jumping to a newly selected focus.

What This Chapter Says About Digital Mind Math

We are adept at grouping details to form concepts, and we have an urge to solidify our understanding of concepts, our knowledge of the details that form the concept and where the concept begins and ends.

Philosophy professor Stephen Yablo sees "aboutness"—an excellent synonym capturing the essence of the boundary of a complete thought—in philosophy, semantics, information theory, library science.[41]

When we satisfactorily identify the boundaries of a concept—when we get a firm handle on what the concept is "about"—we can shift our focus away from the task of understanding the details of that concept. Instead, we now have in our mind a bounded concept, which can be used as a whole within a matrix of interconnected concepts, and from which we can branch off to other related concepts.

The urge to identify the scope of a thought, the boundary of a concept, is a natural and effective tendency. A workably bounded thought embeds a point within our informational matrix and allows us to move on.

Creating the boundaries of a concept helps us maximize information content.

17. The High-P Personal Style vs. the Urge to Adjust the Informational Framework

THE MECHANICS OF DIGITAL MIND MATH

THE ORGANIZATION OF THE MIND

> • **Grouping details to form concepts**
> • **Linking concepts**

• $M^4_+ \times CP_2$ structure: Spacetime 4-spheres, intricately linked

> • **P-adic structure: Sequences of numbered 4-spheres and their connections**
>
> **IN THIS CHAPTER:**
>
> In Digital Mind Math, p-adic numbers label or represent information content. Each p-adic digit labels a concept or an event or a thought. A sequence of concepts is labeled by a sequence of p-adic digits. This sequence of p-adic digits is a p-adic number.
>
> In Digital Mind Math, the size of a p-adic number indicates the magnitude of the information content that the p-adic number represents. P-adic size is measured differently than our normal real size is measured, and p-adic size can increase in two ways: The number of details that each concept comprises can increase, which increases p. Or the number of sequenced conceptual levels can increase, resulting in more digits "to the right of the decimal point." In p-adic mathematics, n is used to indicate the location of the last digit to the right. P-adic numbers increase when there are more digits to the right, which makes n a larger negative number—that is, which increases negative-n.

We'll proceed by examining examples typical of high-p activity, and also examples typical of high-negative-n activity and other activities that adjust the informational framework. All these activities reflect the urge to increase information content.

THE CORE COGNITIVE QUANTUM PROCESS
- Possibility generation
- Selection
- Experience

A Mile Wide and an Inch Deep; A Mile Deep and an Inch Wide

There are two kinds of people in this world.

Well, not really. But let's just imagine two different personal styles of thinking, of conceptualizing, of interacting with others. Let's imagine two different approaches to achieving a goal of digging a hole.

One style examines every detail before making a decision. A mile wide and an inch deep, this Mile Wide personal style is patient and thorough. When you have a big project that you're planning well in advance, this is the person to go to. Mile Wide will examine all the possibilities and set out the advantages and disadvantages of each. Stones will not be left unturned: We'll find out all possible ways of proceeding with every step—any approach that's ever been tried, and some that haven't even yet been attempted. Perhaps, though, Mile Wide may need a little prodding or assistance to make a decision and get going on implementing the project.

The other style jumps to conclusions based on very little data or information. A mile deep and an inch wide, this Mile Deep personal style has a bias for action and is confident about how to proceed based on intuitive leaps but little researched data. Mile Deep is here to dig that hole. Mile Deep sees the big picture—digging that hole— and remains focused on it. Mile Deep draws conclusions fast. Perhaps, though, Mile Deep is a bit impulsive—digs too fast. Mile Deep acts decisively but can leave those around a bit uncomfortable that the quick decision may not be the correct decision.

No, there aren't just two kinds of people in this world. But you probably know some people with characteristics of either the Mile Wide or the Mile Deep personal style.

The Mile Wide style is, in Digital Mind Math terms, the style of high p. In this style of thinking or acting, details of concepts are accumulated, and then accumulated more. As each cognitive moment proceeds, and as there is a choice between gaining additional information or data, versus drawing a conclusion, the high-p choice is the choice that proceeds by gaining additional data.

Note that either choice—additional data, or drawing a conclusion—increases information content. With either choice, recalling Piaget's terms, we are tending toward a higher level of equilibration.

The high-p choice—the Mile Wide choice—is to to increase information content by adding information to existing concepts. This is Piaget's assimiliation: When we are exposed to a new piece of information, we absorb it into our existing informational structure as an additional detail added somewhere in the existing informational structure as one more bits of information. We assimilate the new piece of information into our existing informational framework, rather than disrupt and revise our existing informational framework.

In terms of p-adic mathematics, which in Digital Mind Math is the natural mathematics of cognition, the Mile Wide choice—Piaget's assimilation—increases p. P tells us how many bits of information are included in each level of thought or concept.

For example, before coming across a new piece of

information, suppose Mile Wide had structured the project to have two issues to resolve, with three choices for each issue, each level of the conceptual framework—a first choice, a second choice, or do nothing at this level (skip this level).

With this informational structure in place for Mile Wide, Mile Wide now uncovers a new piece of information. Mile Wide, following Mile Wide's typically high-p personal style, incorporates (assimilates) this new piece of information somewhere into Mile Wide's existing informational structure, rather than modifying (accommodating) the existing informational structure in some way. Let's suppose Mile Wide assimilates this new piece of information by creating a new fourth choice for the second level of decision-making, for the second conceptual level.

In this example, before finding this new piece of information, Mile Wide's conceptual structure envisioned following possible paths with three choices (0, 1, or 2) for the first conceptual step, and three choices (0, 1, or 2) for the second conceptual step.

That is, before finding this new piece of information, Mile Wide had understood the project as consisting of two sequential steps, for each of which there were three possible choices.

Now, with the new piece of information, Mile Wide has a fourth choice for the second step.

Arithmetically, the p-adic representaion in Mile Wide's mind of how the project could proceed was, before the new piece of information, by any 3-adic number from 0.00 through 0.22:

0.00: Skip (omit) both steps

0.01: Skip the first step, but proceed with the second step's Option 1

0.02

0.10: Proceed with the first step's Option 1, then skip the second step

0.11: Proceed with the first step's Option 1, then the second step's Option 1

0.12

0.20: Proceed with the first step's Option 2, then skip the second step

0.21

0.22: Proceed with the first step's Option 2, then the second step's Option 2

The above (with some obvious labeling omitted to make it easier on the reader's eye and brain) was Mile Wide's informational structure before discovering the new piece of information, this new piece of information being that there is an Option 3 for the second step. With the new piece of information, Mile Wide has introduced additional possibilities—0.03, 0.13, and 0.23. With the new piece of information, Option 3 has been added as a possible direction to proceed at the second step.

A few reminders are in order at this point:

- In p-adic mathematics, p is the prime number that tells us how many choices of digits we have. When we can choose numbers from 0.00 through 0.22, we have three choices (0, 1, or 2) for each digit, so p = 3.
- Everything else being equal, the higher the value of p, the greater the size (norm) of a p-adic number, and therefore— in Digital Mind Math—the greater the information content.
- When Mile Wide's choices expand so that Mile Wide may choose p-adic numbers between 0.00 and 0.23, instead of from 0.00 through 0.22, suddenly we can't get by with p as low as 3. P can be 3 when the choices for digits are 0, 1, or 2. But when the choices expand to 0, 1, 2, or 3, that's four choices, not just three. So p—which tells us how many

possible choices there are for each digit—now increases from p = 3 to p = 5 (remember that p-adic numbers require p to take on the value of prime numbers only, and 5 is the next higher prime number after 3).

- Remember the mathematical elegance of universal Witt vectors, also known as big Witt vectors. Although the detailed mathematics of universal Witt vectors is beyond the scope of this book, what's important to know is that mathematicians, through the mechanism of universal Witt vectors, have shown that we may have different values of p for different digits of a p-adic number, and ultimately the arithmetic is the same as if we used the same value of p for every digit. So we—who are proceeding based only on an intuitive sense of p-adic mathematics—do not need to worry that p = 3 for Option 1, but p = 5 for Option 2. We just need to know that a p-adic number with p = 5 is larger (everything else held constant) than a p-adic number with p = 3. (As you proceed with your Ph.D. studies in mathematics, you will proceed with the ghost polynomials and other techniques required for universal Witt vector arithmetic.)

- Also in p-adic mathematics, n indicates how many digits to the right of the decimal point is the last digit to the right. The value of n is positive if there are no digits to the right of the decimal point; for example, if our choices are only whole numbers of 100s, such as 300 or 5200 or 1,823,600, then n = 2. If our choices are numbers from 0.00 through 0.22, then n = -2.

- We are a lot more interested in negative n's than we are in positive n's. That is, it's more interesting when we add levels of detail. For this reason, it's natural to think of "negative n" (-n) as a single concept, more interesting than the concept n itself. Therefore, we'll most often discuss how big negative-n (-n) is, rather than how big n itself is. Concepts are bigger concepts when negative-n is bigger. A p-adic number for which -n = 3 (which is equivalent to n = -3) is a larger concept (has higher informational content) than a concept for which -n = 2.

- Concepts that are represented by a p-adically larger number have greater informational content than concepts that are represented by a p-adically smaller number.
- The size of a p-adic number is given by the calculation p^{-n}.
- Examining the calculation p^{-n}, we note that the size of a p-adic number increases either when p increases, or when negative-n increases. (Incidentally, this is why we said earlier that -n, rather than n itself, is the more basic concept.)
- When Mile Wide assimilates into the informational structure of his project the additional possible choice—choice number 3—for the second step of his process, he has acted in accordance with Topological Geometrodynamics' negentropy maximization principle, which in Digital Mind Math terminology is the tendency to increase information content. Before adding choice number 3 as a choice for his second step, Mile Wide has the choices of 0, 1, or 2 (that is p = 3) for each of his 2 steps (n = 2). Thus the p-adic size of his conceptual framework had been $p^{-n} = 3^2 = 9$. After assimilating choice number 3 into his informational framework as a possible choice for his second step, Mile Wide has increased the p-adic size (which measures the information content) to $p^{-n} = 5^2 = 25$.
- The other approach for increasing information content (this is the approach that Mile Deep takes, which we will discuss next) is to increase -n. As we will see, Mile Deep will, like

Mile Wide, act to increase information content from the same starting point of $p^{-n} = 3^2 = 9$. But Mile Deep—plowing ahead to dig that hole—will do this by increasing -n to 3, so that information content will increase to $p^{-n} = 3^3 = 27$.

- Note that throughout this discussion we have implicitly taken advantage of a feature of p-adic mathematics that we are now going to make explicit: A p-adic number is an ordered sequence of digits, just like a real number is. The order of the digits matters. In Digital Mind Math, when we use a p-adic number to represent a thought, we are proceeding in sequence from the overarching concept— which is the the leftmost digit and is the biggest in a real sense (hence the term "overarching," which sounds biggest)—toward the right, toward the rightmost level of the p-adic number, which is more detailed (smaller in a real sense, but larger in a p-adic sense and consequenty having more information content). In other words, when we use a p-adic number to represent a thought in Digital Mind Math, the sequence of the p-adic digits, from left to right, is entirely critical, and represents the conceptual sequence of the thought, from its overarching placement within our categorization of information, to its more and more detailed and specific placement within our informational structure.

Now on to Mile Deep.

Mile Deep is about to be presented with a new piece of information. Before this new information, Mile Deep's informational structure was set up just as Mile Wide's had been before Mile Wide was presented with the new information: For both, before the new information is available, their informational structures both consist of two choices (plus the choice to do nothing) at Step 1, and two choices (plus the choice to do nothing) at Step 2.

Then our new fact emerges.

Like Mile Wide, Mile Deep is driven to incorporate the new fact into his informational structure in a way that increases information content, in a way that (in Piaget's theory of cognitive development) moves toward a higher level of equilibration, in a way that (in Pitkänen's Topological Geometrodynamics, which is the theory of physics that is the basis for Digital Mind Math) maximizes negentropy or negative entropy, in a way that (in p-adic mathematics' calculation of the norm or the size) increases the p-adic size.

But Mile Deep's personal style is to drive ahead, not to add detail or complication. Mile Deep does not understand this new information as an additional option for Step 2. Mile Deep does not assimilate the new information into his existing conceptual framework.

For Mile Deep, the new information disrupts his existing informational structure. A revised informational structure accommodates (in Piaget's terms) the new information. The new information is enough forMile Deep to plow ahead to the next step.

With the new information, Mile Deep sees the resolution of the first two steps, and is now ready to proceed with a third step.

Mile Deep digs ahead, making a choice at Step 1, making a choice at Step 2, moving on to focus his attention at a new Step 3.

Perhaps what Mile Deep does, upon being presented with the new information, is to make a decision to realize Option 2 at Step 1, to realize Option 1 at Step 2, and then to see that his next task is to decide what to do at Step 3.

Mile Deep realizes the event labeled 0.21—digs those two mine shafts—and finds himself with an informational structure one level deeper, prepared for a new piece of information that will guide him along the higher informational path of Step 3.

The new information will help Mile Deep choose. Mile Deep's new informational structure has three levels—three p-adic digits to the right of the decimal point. His focus is now on what to do at Step 3.

As we've calculated, in this particular case the informational magnitude (p-adic norm) of Mile Deep's choice to accommodate is 27, slightly larger than the informational magnitude (25) of Mile Wide's choice to assimilate. This relative relationship between assimilation and accommodation does not always hold—depending on the circumstances, increasing p can increase p^{-n} faster or more slowly than increasing negative-n. But let's hope that the illustration offers some insight into the p-adic mathematics of Digital Mind Math.

A Meta-Aside

This may very well all seem impossibly complicated. Perhaps this incidental comment—this aside—can give you some comfort or some encouragement.

This isn't just any aside, it's a meta-aside, an aside that looks at itself, an aside that uses the concepts being discussed to discuss the concepts.

This is an aside from the point of view of the author, who sees the concepts being discussed as a single whole concept, an intuitive bubble that is present all at once.

In carrying out the author's burden—writing out the concept word by word in a way that someone else can understand—the author is amazed by how difficult it is to convey his intuition, by how many words and sentences and paragraphs it takes.

After thinking for years about Digital Mind Math and how it works, the author has so internalized it that it is now a single concept, a single thought.

Yes. This is a complicated thought, as demonstrated by how many words and sentences and paragraphs it takes to describe it.

But the fact that it is a single thought—a single p-adic number!—is indicative of the power of p-adic mathematics, the vast informational space in which p-adic mathematics lives.

For Digital Mind Math to be a single p-adic thought requires all of the complexity of p-adic mathematics:

We cannot just have a predetermined set of digital placeholders—2's, 4's, 8's, 16's; or 3's, 9's, 27's, 81's—we must permit fractional exponents between any whole-number exponents. And after settling on certain fractional exponents we must permit more fractional exponents between them, and more between them.

And actually, for the digits of the p-adic numbers, we can't even use normal numbers like 1, 2, 3, 4, 5. We have to use these strange Teichmüller elements, which are the key to Witt vectors, which—in the de Rham-Witt complex's crystalline cohomology—allow whole sequences of p-adic numbers to operate just like a single p-adic number.

And we don't have to worry that one concept has just 2 details, and another has 5, and another has 31. Universal Witt vectors and the universal de Rham-Witt complex allow any combination of p's to act just as though they were all the same p.

P-adic mathematics is so flexible, and p-adic mathematics operates in such a vast informational space, that it is possible for a single fully expressed p-adic number to represent a concept as complex as all of Digital Mind Math.

Now, remember also: In Digital Mind Math, this extraordinarily complex p-adic number is not an end in its own right; it is the mind's natural numbering system that labels the intricately connected M^4_+ spacetime 4-spheres that are our actual, real memories themselves, our little video snippets that, if selected, we can run in our mind as real, experienced memories. Or we can type as real, experienced words and sentences.

Who's Right—High-P or High-Negative-N?

Both routes of thinking and analysis—increasing p and

increasing negative-n—increase information content. Our simplified example above showed the high-p choice increasing the informational quantification to size 25, whereas the high-negative-n choice increased the informational quantification to size 27. But other examples, or tweaks to the above example. would produce different results, some with the high-p choice resulting in higher informational content, some with the high-negative-n choice resulting in higher informational content.

In Digital Mind Math, the experience of the previous thought, along with the experience of external stimuli or new information, set the stage for the core cognitive process to begin anew—a new round of possibility generation, thought selection, and experience.

Both the generation of possible next thoughts, and the selection of the next thought to experience, depend on a number of factors—the structure of our informational framework as it has derived from our entire past history, the focus that the previous thought that we experienced has left us with, the new information obtained from the environment.

If we incorporate new information into our existing informational structure, is this the right thing to do? Is it the most effective way to proceed?

Perhaps the new information perfectly fits the existing informational structure. For example, a child knows that cocker spaniels and beagles are both dogs, and the child is presented with a new external object—a St. Bernard. Once the child understands that a St. Bernard is also a dog, the child has enriched her life (increased the information content in her informational structure) by increasing the breadth of the concept dog from $p = 3$ (choices 0, 1, or 2) to $p = 5$ (choices 0, 1, 2, or 3, forcing the use of the next higher prime number, 5). Here the child is incorporating the newly discovered St. Bernard concept into the dog concept that was previously part of her informational structure, making that informational structure richer by placing another example of the dog concept alongside the examples that had already been present.

And this choice—to add St. Bernard as a third example of the category dog—not only directly increases information content, but also makes the next thought generation process more efficient. By assimilating the concept of St. Bernard into the concept of dog, rather than structuring the concept of St. Bernard as a concept distinct from a dog, the child will be able to generate possible next thoughts and weigh future selection choices more efficiently, since she will be able to use a single concept of dog to include the St. Bernard type of dog, rather than weigh concepts for dogs and also weigh concepts for a separate St. Bernard concept.

So when is assimilation not the right choice?

When would we be better off changing our informational structure in response to new information, rather than absorbing the new information into our existing informational framework?

When would we be better off accommodating our informational structure to reflect the new information, rather than assimilating the new information into our existing informational structure?

First of all, it is important to recognize that some choices—accommodate or assimilate—are neither right nor wrong, but are instead a matter of personal preference, a result of tendencies that emerge from our entire past history of informational choices, from the informational structure and all its weightings as they exist up until the moment of the latest new information. A preferred route for one person may well differ from another's preferred route.

There are, however, cases where it is simply wrong to assimilate the new information into our existing informational framework. A zebra is not a dog. Nevertheless, there are times when we misinterpret events and draw a wrong conclusion, think that a zebra is a dog.

Or perhaps we're just not ready to be able to see that a St. Bernard should be assimilated into the dog category.

In any case, there are times when we accommodate our informational structure to reflect new information, rather than assimilate the new information into the existing structure. There are times when we modify our preexisting informational structure to reflect new information, rather than absorb the new information into the preexisting informational structure.

In our discussion earlier, Mile Deep accommodated. Mile Deep changed the preexisting informational structure to proceed along Step 1 Option2 and Step 2 Option 1, then Mile Deep changed the preexisting informational structure to introduce a new Step 3 that was not previously contemplated, was not part of the preexisting informational structure.

If Mile Deep saw the St. Bernard, perhaps Mile Deep would not have proceeded along the slow, systematic route of adding St. Bernards as another example of dog. Perhaps Mile Deep—with his innate bias toward action, which he translates to an innate bias to increase negative-n—decides to drill down to an additional level of detail, recognizing the need to introduce the concept of groups of dogs: herding, sporting, hounds, and so on. Seeing a St. Bernard, so different from a cocker spaniel or a beagle, has triggered the need to create another level of detail that was not in Mile Deep's preexisting informational structure.

Here accommodation by increasing negative-n was the next correct choice for Mile Deep, because it instantly increased information content by increasing the p-adic norm p^{-n}, which measures information content.

Accommodation can take other forms also. Accommodation is any form of modifying our preexisting informational structure to reflect new information.

Perhaps seeing the St. Bernard results in our next thought

being about how much we love large animals, and how we should download that application that we've been considering to apply to veterinary school to become a large-animal vet. If this is our next thought, then it has emerged from our mind's tendency toward higher information content along with our mind's analysis that thinking about becoming a large-animal veterinarian offers a rich informational path.

To jump from seeing a St. Bernard to downloading the veterinary school application is not a cognitive process of assimilation; it is not a cognitive process whereby a new piece of information is incorporated into our current informational structure. It is definitely a cognitive process of accommodation, whereby the informational structure has changed in response to new information.

In this case—downloading the application for veterinary school— Mile Deep's form of accommodation is not simply in the form of increasing negative-n. This case is a different form of accommodation—still driven by the tendency toward increasing information content, and still involving a changed informational structure (accommodation) rather than incorporation (assimilation) into the existing informational structure—but this time accommodating by migrating our focus of thought, through Digital Mind Math's intricately connected CP_2 thought tunnels, to a different thought to experience, one that the mind evaluates as having high information-maximization potential.

Or perhaps seeing the St. Bernard has triggered a completely different next thought. Perhaps that big hairy St. Bernard has made us realize that we're just not a dog person at all. The thought that seeing the St. Bernard has triggered is not to enhance the complexity of our concept of dog by assimilating St. Bernard into our dog concept, thereby increasing p and directly increasing the magnitude p^{-n} of our information content. Nor is it to accommodate our informational structure to the sight of the St. Bernard by introducing a deeper level of analysis of the dog concept, increasing negative-n by conceptualizing the various groups—herding, sporting, hounds, and so on—of dogs, thereby directly increasing the magnitude p^{-n} of our information content. Nor is it to accommodate our informational structure to the sight of the St. Bernard by migrating through the mind's CP_2 tunnels to the connecting thought of pursuing the potentially higher information route of planning our career as a large-animal veterinarian.

No. This time the sight of the St. Bernard has led us to the thought that we have to tell Junior he's not getting a dog.

This thought is yet another example of accommodation, of changing the informational structure to reflect new informational input. Again we are driven toward information maximization (the great potential for a better informational life that derives from not having a large hairy dog around the house), but this time the new information (sight of the St. Bernard) has driven us to a conclusion, to solve a problem that had been unresolved within our informational structure, the problem of Junior's desire for a dog.

Assimilation increases information content by increasing p. Accommodation may increase information content by increasing negative-n, but it may also increase information content by migrating our next thought toward other high-information possibilities, or by cleaning up our informational structure to make our future informational analyses more efficient.

Schemas, equilibration, genetic epistemology. The work of the Swiss scientist Jean Piaget has been highly influential in the emerging understanding of children's cognitive development. But Piaget considered himself to be first a genetic epistemologist rather than a child psychologist.

Epistemology is the study of the nature of knowledge. Genetic epistemology—the core of Piaget's self-described focus—is the study of the development of knowledge. Here the word genetic is used not in its more typical modern sense, having to do with genes and DNA and inherited traits, but rather in a perhaps more old-fashioned sense meaning development—the genesis of an idea, for example.

So the foundations of Piaget's work—which have been used to draw conclusions about child development and education—are also highly relevant to Digital Mind Math's interest in the nature of knowledge and information and how they develop.

Piaget used the term schema to describe informational structure. (The plural of schema is perhaps most correctly schemata, but modern usage also permits schemas as its pluralization. And you wouldn't be far off to go modern all the way and just use the tem scheme instead of schema.)

A child, for example, can incorporate new information into her mind's informational structure or schema by either assimilation of accommodation, by either adding or combining new knowledge within her existing mental structure or schema, or by changing her existing schema to accommodate the new information or knowledge.

There is no single or specific correct choice or path for the child to proceed along. Instead, the child is driven toward higher and higher levels of equilibration—balanced informational structures that are internally consistent and also consistent with external reality—as a result of the back-and-forth interplay between processes of assimilation and accommodation. Equilibration is achieved at a higher level than previously, only to again be disrupted by additional observation and information, triggering more complex or more subtle or deeper discomfort or contradiction, until resolved as a result of another back-and-forth round of assimilating and accommodating processes that bring us to a still higher level of equilibration.

For Piaget, these processes are biological processes: His earliest formulations of his theories derived from studying mollusks in Lake Neuchâtel near his home in Switzerland. Piaget saw these biological processes resonating with cognitive processes and with the nature of knowledge itself.[42]

For these reasons, Piaget's work resonates with Digital Mind Math also.

In Digital Mind Math, the core cognitive process is drawn from the radical theory of modern physics Topological Geometrodynamics (TGD).

All contemporary theories of quantum physics accept what has to be considered a pretty unbelievable concept—that physics' elementary particles proceed along multiple possible paths until they are reduced to a single path by observation or measurement. It is important to remember that this pretty unbelievable concept—that elementary particles don't have a definite location until they are observed or measured—is an entirely standard and fully accepted aspect of modern physics.

One of TGD's radical elements is to take this core quantum process as a process not just at the elementary particle level, but also as a core process at macroscopic and macrotemporal levels, and a core process of cosmology, biophysics, and cognition.

Digital Mind Math explores TGD's application to cognition, and sees standard quantum physics' core elementary particle process as the core cognitive process—possibility generation; selection guided by information maximization; real experience of the selected thought.

Piaget's primary psychological focus was on the child's cognitive development, from the earliest sensorimotor stage, through preoperational and concrete operational stages, on to the formal operations that continue through adult life. But the processes that Piaget observed in children's cognitive development were also interesting to Piaget the epistemologist: Piaget drew conclusions about the nature of knowledge by studying how a child learns.

Thus it is not a surprise that Digital Mind Math processes resonate so well with Piaget's concepts of schemas, assimilation, accommodation, equilibration.

And as a theory derived from a theory of quantum physics, Digital Mind Math brings in a very broad sense of how new information is brought into the cognitive process to trigger the information-maximizing selection of a specific real thought to experience, selected from a configuration space of many possible choices. In Digital Mind Math, new information comes at us in many forms—not just from observing new information in our external environment, but also from thinking itself, experiencing a thought. The experience of thinking the last thought, just as the experience of seeing a St. Bernard, results in a new round of possibilities for assimilation or accommodation, guided by our quest toward a higher level of equilibration.

Let's continue the quest by more examples of how Digital Mind Math works.

Men Are from Mars, Women Are from Venus

The two planetary categories Mars and Venus—from John Gray's 1993 bestseller *Men Are from Mars, Women Are from Venus*[43]-- illustrate elements of the archetypal tension between the high-p tendency and the tendency to adjust the informational framework.

Here I am not taking a position on how strongly the planets and the genders are correlated—I know this is a controversial point, how different if at all men and women are, biologically or socially. Some men are from Venus and some women are from Mars. And both men and women at times exhibit both Mars-like and Venus-like characteristics.

So let's set aside the debate about whether or not it is men that are from Mars and women that are from Venus. Let's just examine what Dr. Gray's book puts forth as Mars and Venus characteristics and behaviors—regardless of whether they're men's or women's behaviors—and use the Mars and Venus archetypes to continue our illustrations of behavior tending toward high p and behavior tending toward high negative-n and other adjustments of the informational framework.

Solving problems vs. just talking. For example, when Venus wants to talk about a problem, Mars wants to solve the problem, whereas Venus just wants to talk

Both Mars and Venus want to increase information content, but in different ways.

Mars's urge to increase information content is met by adjusting the informational framework in order to find a solution to what Mars frames as the problem under discussion.

Mars listens to Venus (well, kind of listens), but mostly what Mars hears is an enumeration of issues that are crying out for resolution.

Mars is not content to further proliferate details of the problem. Mars has a different path in mind.

Mars envisions adjustments to the informational framework that first involve reorganizing, regrouping, and renaming various issues that Venus has enumerated. Mars is very good at this. Mars has had lots of practice simplifying issues to a common core that Mars is familiar with.

Venus may feel that Mars is pulling back the conversation to a smaller negative-n—that Mars is removing the focus from details, many emotion-laden, and instead is refocusing to a more general topic that appears to Venus to have lower information content because it has reduced the number of levels of detail.

Mars is guilty as charged. Mars is attempting to move the discussion away from a level with more details toward a more summary level in which we have a lower negative-n and therefore a (temporarily) lower information content.

This is a source of high tension, a double dose of

dissatisfaction for Venus. First of all, Venus wanted to continue increasing information content by continuing to enumerate details, resulting in higher and higher p for the topic under discussion. Mars has signaled that p is high enough—we've been presented with enough details—and that it is time to bring the topic to a level with fewer layers of detail; it is time to introduce a group name or a grouped conceptualization of these details, a smaller negative-n. So not only is Mars signalling a move away from increasing information content by continuing to increase p, in addition Mars is exhibiting a second behavior dissatisfying to Venus: The particular move away from Venus's preferred approach is a move that decreases information content, that moves to a lower negative-n.

Venus is furious.

The disconnect is that Mars has a plan. Mars's personal style urges toward adjusting the informational framework. Mars sees a path ahead, which admittedly includes some temporary short-term setbacks, but which will ultimately lead to a higher informational state.

Accommodation—Piaget's label for Mars's approach—can be messy. Accommodation—changing the informational framework to reflect new information—can result in an informational framework with a temporary decrease in information content.

In the back-and-forth path toward a higher level of equilibration that Mars envisions. the first accommodation implemented with respect to the informational framework is to make Mars-sense of the details that Venus is enumerating, by beginning to group them in categories that will allow a different informational structure.

Mars, the informational framework adjuster, sees this process as a whole, one that can proceed rapidly, but one that will have its temporary setbacks as it lurches back and forth, accommodating and assilmilating, toward Mars's vision of a more stable equilibrium.

Enumerated issue by enumerated issue, Mars intends to group some (accommodating the informational structure to a lower negative-n, so that several issues or details can be treated as one), and even intends to align others (assimilate!) within Mars's preexisting categories, because to Mars they look alike, appear as though they are an additional example of a phenomenon already labeled and stored within Mars's preexisting informational structure.

This is all very frustrating for Venus, who prefers to continue to increase information content by clean, straightforward, direct assimilation. Increasing p—increasing the details shared about the topic at hand—directly increases p^{-n}. This is Venus's way to go, and Venus does not see the need to adjust the informational framework, particularly when Mars's approach not only stops in its tracks the continuing increase of p, but in addition, initially at least, decreases negative-n, moves negative-n in the wrong direction, makes negative-n a smaller negative number, thereby reducing informational content p^{-n}.

Mars is in a state of disequilibrium. Not sharing Venus's high-p personal style, Mars's mind is ripe for adjustment of the informational framework. And Mars is good at it.

Mars groups details in unexpected ways—increases p within existing and new informational categories. Mars regroups these groups—changes the subject!—and, given the chance, reanalyzes (digs deeper, increases negative-n) with respect to this regrouped informational framework.

Mars's well-developed sense of adjusting the informational framework allows Mars to see in an instant a map of where Mars is headed—a streamlined informational framework that allows efficient generation and selection of possible next thoughts, and a state of equilibrium.

Perhaps this is a state that Venus will enjoy, that Venus will share with Mars as a higher level of equilibration due to its new perspective on the issues or due to its resolution of the problems at hand.

Or perhaps this is only Mars's state of equilibration, in which Mars is done and can go back to watching the important football game. Venus will telephone Venus2 and continue increasing p.

Both Venus and Mars are driven to increase information content, but in different ways. Venus is exhibiting a high-p personal style, whereas Mars is exhibiting the urge to adjust the informational framework.

Retreating to the cave. When faced with stress, Mars may want to escape to another topic entirely, perhaps in the Mars cave, watching sports games, lost in another world.

Here Mars chooses against looking to the stress-inducing topic itself as the source for increasing information content. Mars chooses to discontinue discussion of the topic at hand. Instead Mars jumps to another topic, one offering Mars more promise for increasing information content: an existing hobby, for example, which offers promise for increasing information content by adding details, or by broadening the understanding.

This, in Piaget's framework, is for Mars another form of accommodation, another way in which Mars's informational structure adapts to new information not by assimilating it into the current informational structure but by changing the structure.

When Mars retreats to the Mars cave, he most definitely is not assimilating. He is not taking in the new information offered by Venus and adding the new information as additional detail within the topic at hand. He is not increasing p as his way of increasing information content.

Mars is definitely accommodating, in the sense that his informational structure has changed. But he has done this by moving his focus to an entirely new place, not by pushing forward the resolution of the details of the topic under discussion.

Yet Mars is still driven to reach a higher level of equilibration, to achieve greater informational content. Piaget's accommodation has more varied manifestations than Piaget's assimilation, which always takes the form of incorporating a new detail as an additional point of information within the existing informational structure.

Analyst Christopher Jones presents a useful metaphor: "Assimilation is like adding air into a balloon." At every step, you keep blowing air into the balloon, and it gets bigger and bigger. Accommodation is like twisting the balloon into a new shape, like an amusement park performer shaping a round balloon into a poodle.[44]

In this case, Mars has reached his limit (at least his self-perceived limit) for how far he can take his current conversation with Venus. He can't assimilate anymore, and he can't even proceed with the form of accommodation driving toward a proposed resolution of the topic at hand. Instead he escapes to another topic.

The mind is very big place, and Mars has other things going on in his mind besides the current topic under discussion with Venus (as important as it is). At times, the urge to adjust the informational framework is manifested by actions that directly increase information content p^{-n} by directly increasing negative-n: for example, asking Venus for more details of an aspect that she is discussing. At other times, the urge to adjust the informational structure is manifested by acting to clean up the informational structure in order to make more efficient the cognitive process of generating possible next thoughts or actions.

But for this moment for Mars, it's the Mars cave. A change in focus, most likely to Mars's own little sphere, much less important than the discussion with Venus that he is escaping from. But Mars will return, energized by the increase in information content in the Mars sphere, with new energy for possibility generation, selection, and experience with Venus.

In the meantime, while Mars in the the Mars cave, Venus can continue increasing information content according to her preferred high-p style by finding a fellow Venus friend to talk with. Venus 1 and Venus 2 can go on for hours engaging in Venus talk, attentive to each other, listening, sharing more and more details, with no pressure to judge or to find a resolution that closes the subject, and with no jumping from topic to topic, away from the topic at hand.

Ultrametricism. Dr. Gray, in *Men Are from Mars, Women Are from Venus*, makes a Venus/Mars distinction that resonates strongly with a specific aspect of p-adic mathematics—the ultrametric nature of p-adic numerals, the fact that the 7-adic numeral 1 does not represent a bigger or smaller number than the 7-adic numeral 2 represents or 3 represents. They all represent numbers of the same size. They are just the different names for concepts that are all at the same level of detail and are all the same size.

This ultrametricism is, in *Men Are from Mars, Women Are from Venus*, a very Venus way of thinking. As Dr. Gray puts forth, Mars has a point system by which small things he does for Venus get a lower number of points, and large time-consuming or expensive or difficult things he does for Venus are evaluated with more points.

But Venus doesn't see it this way. Venus is ultrametric. Everything gets one point in Venus's scoring system. It's the acts of attentiveness that count, and a high score is obtained in Venus's scoring system by sustained attentive acts, big or small.

There is no trump card in Venus's scoring system. For Venus, this is a sustained relationship that grows and grows by the continuing shared activities. This is very consistent with the high-p style of increasing information content, in which the natural style is to stick to the topic at hand and increase information by increasing the details of this topic.

To keep focused on the topic at hand, it is critical to smoothly incorporate additional details without extensive evaluation of the new details. We're discussing a topic. Stick with it. Don't shine a very bright light on a new detail, because inevitably that would change the focus to a new topic, a distracting accommodation that is contrary to the high-p assimilation that is Venus's preferred personal style.

The high-p style focuses on sharing information without evaluative intent, without measurement, without metricization. The high-negative-n style, and other forms of adjusting the informational framework, focus on getting to the point, on solving a problem, on not letting the perfect be the enemy of the good, on making the informational process more efficient.

The Single Story

Nigerian novelist Chimamanda Ngozi Adichie discusses stereotyping in her 2009 TED Talk, "The Danger of a Single Story." Adichie discusses the temptation for majority or powerful groups to immediately pigeonhole a minority or less powerful person as that minority, nothing but that minority, with all the stereotypical characteristics of that minority, and without characteristics common to the members of the majority society. Adichie warns of what is lost by this "single story" orientation, and she asks us to consider another approach:

> "Our lives, our cultures, are composed of many overlapping stories. . . If we hear only a single story about another person or country, we risk a critical misunderstanding." [45]

P-adically, through Digital Mind Math, we may understand the recognition of others' many stories as an action increasing information content by increasing negative-n: creating an additional conceptual level in which we will record the stories and details that we learn about another's life. In Piaget's terms, this is accommodating behavior: We accommodate our preexisting informational structure to reflect new information, in this case changing the preexisting informational structure by opening up a new level in which we will store the details of our new acquaintance's personal story.

On the other hand, we may p-adically understand the relegation of others to a single story as a structurally static action, keeping the informational structure intact and constant, although increasing p as we now have met yet another person matching the traits of those we have already grouped within this single story.

And we can also, as a third possibility, have a different thought entirely, thinking about the meaning of our urge to force the categorization into single stories, perhaps leading to new understandings of ourself and the way we think and act. This third possibility is not the simple accommodation of our first possibility, in which we go one level deeper. Nor is it the assimilation of our second possibility, in which the new experience is absorbed within our existing informational structure. Rather, this third possibility is a more complex form of accommodation, in which our focus has found a connection—through a CP_2 wormhole—to a related concept from which we will continue our thoughtful experience.

Let's look more closely.

When we encounter new information (meeting a new person, for example), we generate possibilities about the role, if any, that this new person will play within our next experience.

One possible next action is to focus on this new person with a plan to proceed to add information about the new person. That is, first we act to open up a new 4-sphere at a more detailed level (higher negative-n) than our preexisting level within the people category. We reorganize our concept of this new acquaintance, away from being just one more example within our prexisting single story 4-sphere, and instead increase negative-n by opening up the new topic. Then we populate this new 4-sphere—this new topic—with individual characteristics that are not simply the preconceived characteristics of the single story. We are opening up the possibility that there will be future experiences that will add more information about this new person, this newly opened up 4-sphere in our mind.

So we first accommodate our informational structure in order to open up a new focus, then we will assimilate additional detalis at the level of this new focus, increasing the bits of information we have about this new person, increasing our information content by increasing p, the number of data points that, in our mind, define who this person is—doing this, that is, as the next step after having first increased negative-n by opening up a new 4-sphere.

After opening up our informational structure to create a new level of information for our new acquaintance, our next experience after that will keep our focus on this new person, and will proceed by adding information about this person as a person, in this person's own terms. The consequence is that we first increase information by increasing negative-n, then we continue to increase information by increasing p, always ready to generate possibilities for a next experience with a new starting point of the just-enriched conceptualization of this recently met person, for whom only a few moments ago we established her own personal M^4_+ spacetime 4-sphere.

This is in contrast to a different choice as the first experience upon observing this new person, a choice not to focus on this person at all, not to increase our level of negative-n by creating a new more detailed category of person for this new acquaintance, but instead to incorporate or reduce the experience of this new aquaintance entirely within our preexisting informational structure. Perhaps we categorize this person's traits and actions as fully described by labels and categories that already exist in our mind's informational structure. We completely assimilate our entire experience of this new person into our existing informational structure, identifying her completely within a preexisting single story.

Now let's not get all Pollyannish about this. We all have busy days, and none of us can proceed by creating a focus on every new person we see and finding out more about that person in that person's own terms. For most of our day, we must make choices to continue to focus on the day's planned activities, and not drop that focus every time we see a new person to find out that new person's story.

This does have the effect of streamlining our informational structure, without too many loose-hanging barely connected concepts, permitting us to get on with our thought-work with a clean structure that does not have bothersome extra negative-n levels with just one element in it, which would make future generations of possibile next thoughts messy and more complex.

But suppose you're flipping TV channels and you see the Nigerian Ibo novelist Chimamanda Ngozi Adichie talking about the danger of a single story.

You hear Adichie saying, "The single story creates stereotypes, and the problem with stereotypes is not that they are untrue, but that they are incomplete."[46]

You might now generate as a possible next thought the possibility of retaining the focus on Adichie to see and hear more from her. This would be similar to the increasing-p possibility that we've already discussed, creating additional detail about a newly created concept at a more detailed level, a concept which we have just opened up at a level of higher negative-n when we retained our focus on Adichie.

If you postponed flipping the channel past Adichie for a moment or two, and experienced what she had to say, then you first increased your level of negative-n, creating "Chimamanda Ngozi Adichie" as a new more detailed level of your "people" concept, and then you increased the value of p associated with the concept "Chimamanda Ngozi Adichie" in your mind's informational structure. There now is a labeled Adichie concept in your mind, and in future generations of possible next thoughts, you will come across Adichie, and she will be linked to the meaning of "the single story," and maybe this will affect the next thought you select to experience.

Or you might have yet a different experience upon hearing Adichie say that stereotypes, although incomplete, are not necessarily untrue. Hearing this may shift the focus somewhere else in your informational structure, branching off in a different direction after experiencing Adichie's comment about the single-story stereotype. Maybe hearing Adichie say that, although incomplete, stereotypes are not necessarily untrue could get you thinking about a whole set of informational formulations that you have regarding sociological and anthropological categorization, about what we gain and what we lose in our day-to-day construction of single stories.

The privilege to be unique. In a CNN article discussing how coverage of violence perpetrated by white Americans often differs from coverage of violence perpetrated by black Americans, commentator Sally Kohn discusses how one of the characteristics of majority or powerful groups is the "privilege to be unique."[47] Kohn describes the frequent tendency in news coverage for crimes committed by the majority group to either not mention the race of the criminals, or to go out of the way to describe the crime as unusual and not representative of the majority group to which the perpetrator belongs. By contrast, crimes committed by minority groups generally mention the perpetrator's race, and often include a context-setting comment indicating the high-crime nature of the area or community in which the crime took place.

Kohn's analysis is consistent with the "single story" framework, and shows the dichotomy between allowing a level of higher negative-n in which personal details accumulate, vs. incorporating (assimilating) another's personal details into a preexisting single story.

The "privilege to be unique" is a perfect statement of the route that allows individuals their own higher-negative-n level in which their personal details accumulate. Allowing individuals' uniqueness to be examined is exactly what it means to create and retain focus on that individual and to gain additional information about the individual per se, the individual herself.

A simple assimilation—avoiding creating a new personalized category, but instead seeing just an additional entry within a single broader preexisting category— may keep our life streamlined and simple by not creating a discordant one-element extra level of negative-n that makes future thought processes complicated and bothersome.

And a third option—springboarding (through a CP_2 connection), as Kohn does, to a different concept entirely, the concept of how to make sense of what this means about us—may get us to think about the tension between the structural streamlining vs. the incompleteness and lost opportunities that result from stereotyping.

Engineering approximations. Engineers build things, and often can be distinguished from other, more theoretical scientists by engineers' bias for action. Engineers often don't wait for a perfect theoretical explanation or analysis, but instead proceed based on practicalities, often empirical estimates for what will work, even if there is no grand theory to explain why.

Alex Pentland is a data and computer scientist who recently published the book *Social Physics*, about how ideas spread, and about social influence and collective intelligence.[48]

A concept that Pentland uses in this book is how "engineering approximations" can help slice through the mass of big data that is produced online. It's one thing to talk about how engineering approximations can be useful to solve practical problems in building engines or bridges. But it's more striking to apply the term to a process of drawing social implications from large databases.

Engineering approximations within "social physics" are a classic technique of assimilation. They are not a technique that delves deeply into the broader cultural context in which social phenomena exist; they are not a technique looking to understand interconnections or interpretations. They are a technique that examines how data patterns can be simplified to resemble easier-to-use mathematical patterns, and to work with these preexisting mathematical patterns as a proxy for a deeper or interpretive understanding.

Engineering approximations remove the focus from the idea itself in order to fit the idea into a known, established framework.

In Love or Business Relationships, When There Are Differences

The *New York Times* has a regular Sunday Business column called "Corner Office," interviewing business executives about their backgrounds, influences, and approaches to managing and hiring.

In 2015 Lois Braverman was interviewed. She is the Chief Executive Officer of a New York-based family therapy organization, the Ackerman Institute for the Family.

Braverman describes a personal characteristic drawn from her background as a psychotherapist that she has found useful not only in marriages and parent-child relationships, but also in business leadership:

> "If people are in conflict, it's because we really are perceiving some aspect of the world differently at that moment. . . I have to make room for the legitimacy of your point of view, and not let my

righteousness make me think my perception is more meaningful than yours." [49]

Let's examine Braverman's advice to see how it can be understood p-adically.

Braverman suggests that a conflict can derive from differing informational contexts in which the two parties place some aspect of the world.

Both parties see the same aspect or detail, but they have surrounded this detail with two different understandings, two different conceptual frameworks.

In looking to make room for the legitimacy of the other point of view, Braverman is looking to enlarge her own informational structure that surrounds this detail so that, while still preserving a framework meaningful to Braverman, this informational structure is large enough to incorporate the other party's point of view.

Braverman is not saying to herself or out loud: "I have a complete understanding of this aspect of the world. Therefore, we need to address what's wrong with your understanding of the world that leads you to a wrong understanding of this particular aspect."

No. Braverman is not changing the topic to discuss what's wrong with the other person's understanding. She is enlarging her conceptualization of this topic to add a new point about this topic, a point that incorporates the legitimacy of the other person's point of view.

This first step in a situation of conflict—a step to increase p—may or may not lead to an immediate equilibrium. There may well be a need to accommodate informational frameworks, to go back and forth with assimilation and accommodation before a more stable equilibrium is reached. But the first step of making room—increasing p—can be a useful technique in the quest for higher levels of equilibration and information content.

The Loser Edit

If you've spent more time than you should watching a favorite reality TV show or two, you may have noticed how the shows' editors sometimes foreshadow the upcoming loser by the way the edited version offers hints of the problems that the loser is experiencing, the conflicts with others, the failures to bake correctly or string together words coherently or amicably.

For a weekly hour-long reality show, dozens or even hundreds of hours of video are recorded. Pretty much anyone could be shown to be a fascinating character with great prospects, or the next loser.

To the extent that the goal of the show's editors is to make sense of the show for the show's viewers, the editors may be tempted—based on their advance knowledge of who's about to be voted off or otherwise designated as the episode's loser—to edit the large amount of available tape to show a storyline that accomplishes a specific purpose, making sense of why the loser lost.

Of course, at this point, several decades into the reality TV phenomenon, many viewers have become much more sophisticated about the editor's bag of tricks, so editors may want to play with the mind of the informed viewer, making them think that they've figured out all too easily who's about to lose, but it's all been a deception on the part of the editor, a ruse to wake the viewer s from their smugness, a technique of reverse psychology.

This can add a whole layer of interest and excitement. But let's just stick with first-order loser edits, direct loser editing, done for the purpose of having a one-hour episode stand on its own merits as a sensible plot in which characters' motivations are understandable.

The goal to present a sensible hour-long plot with understandable character motivations is a goal to achieve equilibrium at the episode's end. For the episode to make sense for the viewer, the outcome needs to appear as a natural consequence of the information presented. We need to know why the loser lost.

The process of producing the loser edit is a process that reverses our more common approach to life.

As we participate in the normal everyday unfolding of our life, we are faced with externalities that we need to make sense of. Some phenomena are consistent with our preexisting understanding and reinforce that understanding. Other phenomena appear to contradict our preexisting understanding, and can therefore trigger us to dig deeper, find out more, change our understanding of the phenomenon, or modify our previously existing informational structure.

In the normal everyday unfolding of our lives, we do not know in advance how things will turn out. We may well end the day with an informational structure that has been modified from the structure we began the day with.

This is how a reality show's editors want us to feel about the hour we spent watching this week's episode: The episode is intended to allow the viewer to follow along, assimilating and accommodating, sixty minutes later at equilibrium, satisfied that the episode's trajectory has made sense.

But this is not how the editors themselves proceeded. They have reverse-engineered the plotline, so that the ending conclusion is sensible and exciting, and probably so that viewers, momentarily content with equilibrium achieved at minute sixty, will very soon feel the urge to seek a higher level of equilibrium—satisfied with this week's ending, but with a week to build up excitement for reaching a higher equilibrium.

An essay "Life Without Pity," by the novelist Colson Whitehead, was recently published in *The New York Times Magazine*'s First Words column. This essay segued from reality shows' loser edits to an analogous process—"the plausible argument of failure"—that occurs in our real non-televised lives.

Colson discusses the "people we have demoted to the status of mere failure"—the fired, the dumped, and those otherwise mortified—and how we construct our own loser narrative about these unfortunates in our own minds and for discussion with others.

Colson broadens this to how we "clasp meaning onto existence" and onto our own internal narratives that we construct and play back, even if only to ourselves, in our own mind:

> "Memory is the most malicious cutter of all, preserving, recasting, panning in slow motion across the awful bits so that we retain every detail. . . Cut when this scene establishes the perfect pitch. . .Splice and snip. The contradictory evidence falls to the cutting-room floor."[50]

This is p-adic surgery. Digital Mind Math surgery.

This is the selection and editing and ordering of snippets—M^4_+ spacetime 4-spheres, each p-adically labeled and given meaning by its p-adically labeled CP_2 connections to other M^4_+ spacetime 4-spheres—reordered to create a new story, with ordering and linking all captured by a grand p-adic number that labels for us the edit we intended.

What This Chapter Says About Digital Mind Math

Practice makes perfect.

The small examples in this chapter offer practice in various aspects of applying p-adic mathematics to how we think and to how our minds work.

We've practiced with several ways in which we attempt to increase information content—to increase the norm or the size of the p-adic magnitude of our thoughts—by adding detail, by increasing the number of levels of detail, by migrating to connecting information.

The examples are also exercises in the use of p-adic mathematics within the framework of Digital Mind Math, numbering Digital Mind Math's M^4_+ spacetime 4-spheres and their CP_2 connections which form our ever-emerging minds.

In the next chapter, we'll look more at these connections.

18. Portals

THE MECHANICS OF DIGITAL MIND MATH

THE ORGANIZATION OF THE MIND
- Grouping details to form concepts
  - Linking concepts

- $M^4_+ \times CP_2$ structure: Spacetime 4-spheres, intricately linked

IN THIS CHAPTER:
Here we focus in particular on the intricate linking that Digital Mind Math's four-dimensional CP_2 space allows.

In Digital Mind Math, each thought—that is, each M^4_+ spacetime 4-dimensional sphere—is linked in CP_2 space to all thoughts that it is a component of, as well as to all ways in which it can be resolved into components, as well as to all thoughts that share with it a component or theme or subject-area tag or emotional or physical resemblance.

We have noted on numerous occasions that p-adic mathematics is brilliant at straightforwardly recording this mapping, labeling these pathways. But it is not p-adic mathematics that is the subject of this chapter. It is the real CP_2 pathways themselves that are the subject of this chapter.

In this chapter, we envision portals as a model for these pathways.

- P-adic structure

THE CORE COGNITIVE QUANTUM PROCESS
- Possibility generation

- Selection
- Experience

Portal is video game produced by the Valve Corporation, set in the mysterious Aperture Science Laboratories. A key feature of Portal is that players have access to portal guns, which permit the simultaneous creation of two windows on a single flat surface. Movement through these open portals is possible no matter how distant the connection. Players use a handheld portal device to complete a series of puzzles. [51]

Military analysts use the portal gun as something of a metaphor for the ultimate game-changing military technology that would permit instananeous transfer of weapons and troops from friendly bases to enemy locales. For example, in a discussion of "seabasing" strategies, portal guns are the ultimate theoretical concept for solving logistical problems of basing a military force on the other side of the world:

> "A real-life portal gun would be among the most disruptive, destablizing military technologies ever because of its impact on logistics. Arguably, it would be even more disruptive than the development of gunpowder and nuclear weapons.

> "There would be no point in maintaining overseas bases, as distance would basically become irrelevant. Imagine the US Army driving tanks and artillery from a base in Texas directly to a battlefield in Iraq . . ."[52]

Because, for serious military analysis, portal guns are only a metaphorical ideal, rather than a serious practicality, analysts discuss how closely this ideal could be achieved by the remote seabasing of troops and arms, taking advantage of advanced mobile landing platforms as part of a strategy to simplify the establishment of needed military infrastructure.

Portal guns are not a practical military reality, but they're a great description of the CP_2 space in which Digital Mind Math's unlimited interconnections reside.

Digital Mind Math's eight-dimensional $M^4_+ \times CP_2$ space is structured as M^4_+ spacetime 4-spheres intricately interconnected by wormholes that reside in four-dimensional CP_2 space.

CP_2 "complex projective" space has four very small dimensions. That these dimensions are very small—smaler than the resolution available to us through our sensory capabilities—explains why we do not observe these connecting wormholes. We experience the macroscopic and macrotemporal M^4_+ 4-spheres in large-scale four-dimensional Minkowski spacetime. But the four spatial dimensions in which the connecting CP_2 wormholes reside measure only 10^{-30} meters.

It is the fact that CP_2 space is four-dimensional—not three-dimensional—that gives CP_2 space the power to infinitely interconnect or link any of Digital Mind Math's M^4_+ 4-spheres to any number of other M^4_+ 4-spheres, no matter how distant and no matter how many other links already exist. This extra dimension—this fourth dimension—allows unlimited intricate linking of macroscopic and macrotemporal spacetime 4-spheres within Topological Geometrodynamics' $M^4_+ \times CP_2$ eight-dimensional embedding space.

The portals available in Valve's Portal game (and put forth as a military strategist's ideal for solving logistical problems in projecting global power) are a perfect functional vision of the limitless interconnection among M^4_+ 4-spheres that four-dimensional wormholes in Digital Mind Math's CP_2 space create.

What This Chapter Says About Digital Mind Math

We experience each thought as an M^4_+ film clip, whose tags identify it and its components as functionally similar to any number of other M^4_+ film clips, and also identify it as a component of other M^4_+ film clips, and also identify the various combinations of other M^4_+ film clips that it encompasses as its components.

A new $M^4_+ \times CP_2$ mind map—incorporating the latest M^4_+ experience, stored with its intricate CP_2 interconnections—is ready to generate a new set of possibilities for the next M^4_+ thought to experience.

The portal gun technology of CP_2 space permits limitless linkings.

Fortunately for us, p-adic mathematics, the natural mathematics of cognition, operating in its vast p-adic informational space, allows straightforward mapping and remapping of our mind space, generating multitudes of possible next thoughts, and measuring their p-adic sizes to indicate the magnitude of their information content.

A next thought is selected, tending toward maximization of information content.

Our mind jumps through a CP_2 portal to experience that next thought.

Then, from that next thought, we start again, just like the core quantum process of elementary particles such as the electron: possibility generation, selection, experience.

With the CP_2 aspect of Digital Mind Math's $M^4_+ \times CP_2$ space now mastered through the intuition provided by the concept of portals, it is time to gain a firmer intuition for another feature of $M^4_+ \times CP_2$ space, namely the 4 of M^4_+ . This is the fourth dimension—time—of special relativity's four dimensional Minkowski spacetime.

We'll master the concept of time as the fourth dimension through the intuition provided by the concepts in the next chapter.

19. Dogs and Phner: Living the Four-Dimensional Life

THE MECHANICS OF DIGITAL MIND MATH

THE ORGANIZATION OF THE MIND
- Grouping details to form concepts
  - Linking concepts

- $M^4_+ \times CP_2$ structure: Spacetime 4-spheres, intricately linked

IN THIS CHAPTER:
Here we focus in particular on spacetime 4-spheres, especially their fourth dimension, time.

What is it like to think four-dimensionally? How would our thoughts and minds proceed if, as an everyday way of being, we understood the world with the time dimension being a natural a part of our conceptual framework, fully integrated alongside the three dimensions of space?

- P-adic structure: Sequences of numbered 4-spheres and their connections

THE CORE COGNITIVE QUANTUM PROCESS
- Possibility generation
  - Selection
  - Experience

Dogs

Dogs love to sniff as a way of greeting. They sniff fellow dogs, and they sniff people. They appear entirely involved in these first few moments of greeting, a ritual deep in their nature and their way of being.

What is all this sniffing about? What drives dogs to sniff as their ritual greeting?

Here's a hypothesis, maybe only part of the answer, but nevertheless useful to illustrate what it's like to live four-dimensionally, to conceptualize the world not just through the three spatial dimensions which we see at the current moment, but rather to conceptualize in extended time, with the world of three spatial dimensions passing through the fourth dimension, time.

A creature that conceptualizes four-dimensionally does not think via still images, but instead thinks via moving images—film clips, starting at an earlier moment of time and proceeding to a later moment of time.

Actually, this is exactly how we all think, but for some reason the concept of a fourth dimension (time) invokes mystery and science fiction rather than ordinary grounded reality. So let's take a moment to think like a dog, then circle back to see how much this is like how we think.

Dogs have a highly developed sense of smell, more highly developed than humans do: Dogs have over 220 million olfactory receptors, whereas humans have only 5 million.[53] As a result, dogs' orientation to the world around them has an element of reliance on their sense of smell that is greater than humans exhibit.

Smells have a way of attaching themselves to our (and to other animals') bodies, and to our clothing and anything else we have with us. Smells persist. And this is exactly what is meant by traveling through the fourth dimension, time.

When a dog greets its master, after the master's long day away at work, the dog sniffs its master's pant legs as the dog's way of living four-dimensionally.

In the current instant that the dog inhales the odors from a small spot at the bottom of its master's pants, the dog is living through the master's day. The dog is experiencing what happened since the last time it saw the master: Who did the master interact with? What food was he near? What sites did he walk by? What was in the bakery trucks and meat trucks and garbage trucks that drove by?

Each sniff evokes a whole story, a whole set of three-dimensional images evolving over time.

The greeting sniff is a highly efficient way for the dog to catch up on its master's day of activities. Dogs are particularly drawn to this style of greeting because of their highly developed sense of smell, and because of smells' persistence over time.

As the dog lives from moment to moment—a sniff here, a sniff there—in each moment, with each sniff, the dog experiences a moving image, lives through a film clip, conceptualizes three-dimensional objects interacting and moving through the fourth dimension, time.

This dog is living the $M^4_+ \times CP_2$ life. It is living a life where its concepts are M^4_+ 4-spheres—film clips, moving images. And in the dogs mind, the M^4_+ 4-spheres get processed and labeled and connected with other M^4_+ 4-spheres. The dog experiences and continually reorganizes its intricately linked spacetime 4-spheres.

This is how, in Digital Mind Math, we all think, how we all organize the mind.

This may seem mysterious, but it's actually just the natural method of cognition.

It takes a dog—with its heightened sense of smell—to help us remove the mystery from one aspect of the $M^4_+ \times CP_2$ mathematical framework, the aspect of what it means to think using M^4_+ 4-spheres, with time as the natural fourth dimension.

Phner

In the science fiction book *In the Cube*[54], global warming has sunk the city of Boston beneath the sea, but just off the coast is an intergalactic spaceport that receives extraterrestrial travelers from throughout the universe. The main character is a human detective, but her sidekick is Akktri, a Phner extraterrestrial who has arrived through the spaceport portal.

The Phners are beaver-like beings with highly developed senses of smell and also with an orientation to the world very different from the human orientation. The Phner interact with other beings not simply by experiencing the current moment, but rather by experiencing the other being's entire life history from its birth until its death.

A Phner experiences another being, all at once, in the other being's past, present, and future.

As a result, Akktri offers particular advantages as a detective's sidekick. But in addition, the Phner have some very different perspectives than humans do on the meaning of life and the meaning of death.

In a definite sense, a being is already dead when a Phner interacts with it. Death is very much part of life, always present alongside birth and alongside all of life in between. Phners appear to other beings to have ESP-like senses, but these appearances of extrasensory perception are just natural consequences of the Phners' fully four-dimensional way of thought.

I suggest that it is not coincidental that the beaver-like Phners' most highly developed sense is the sense of smell.

The Phners live the the M^4_+ four-dimensional life, each moment an M^4_+ 4-sphere moment, processed and stored in their ever-emerging $M^4_+ \times CP_2$ minds of intricately linked spacetime 4-spheres.

<u>The block universe theory of time</u>. Actually, the Phner

conception of time is essentially what philosophers refer to as the block universe theory of time, in which objects are considered spread out in time, just as they are spread out in space. Time is one of the dimensions of spacetime, part of "the fabric of the universe, not something moving inside it."[55] Objects don't exist in a now-disappeared past or in a not-yet-existing future; "they simply exist in different parts of spacetime."[56] A Phner couldn't have said it better.

Geometric Time and Subjective Time

Topological Geometrodynamics gives us a couple of concepts and terms that are useful here, in our discussion of the fourth dimension, time.

Four dimensional Minkowski spacetime M^4_+ is a specific four-dimensional geometry that emerges from the theory of relativity—from special relativity in particular, the entirely well-established and understood way in which the three dimensions of space and the fourth dimension, time, are affected by motion.

According to special relativity, the three dimensions of space—left/right, up/down, and forward/backward, often referred to as the x-axis, the y-axis, and the z-axis—are at 90° to each other (mutually perpendicular or orthogonal; at right angles) only for objects not in motion. Once an object is in motion relative to a stationary observer, the three spatial dimensions of the object are no longer at perfect right angles. And this phenomenon becomes more pronounced as the relative speed gets greater and greater, closer and closer to the highest possible speed, the speed of light. Special relativity also brings time into this geometry, giving us the four-dimensional geometry of M^4_+ Minkowski spacetime.

Geometric time—in the geometry of four-dimensional M^4_+ Minkowski spacetime—moves only forward, and proceeds according to mathematical specifications of the particular M^4_+ 4-sphere that is under consideration. In the case of Digital Mind Math, each M^4_+ 4-

sphere is a thought bubble, a film segment, in which objects in three spatial dimensions proceed geometrically forward through the fourth dimension, time.

So geometric time is one conception of time, the conception that orders M^4_+ thoughts from their beginning to their end.

Geometric time is not the only time of Digital Mind Math, however. Our mind is organized as sequences of intricately linked M^4_+ spacetime 4-spheres. We think by experiencing ordered sequences of these 4-spheres.

When we think, we proceed along a sequence of M^4_+ 4-spheres in our own subjective time. When we generate possibilities for our next thought, we can reach backward to past experiences and forward to hypothetical future experiences. This is subjective time, the time that we experiment with—backward and forward—as we generate possibilities for the next thought, and the time that we experience in our memories and daydreams and hopes and regrets.

Three-dimensional spatial objects proceed in M^4_+ spacetime through the fourth dimension, geometric time. Our mind sequences these M^4_+ spacetime 4-spheres in a four dimensional CP_2 web, from which, guided by information maximization, we select one particular sequence to experience in subjective time.

In philosophy and in physics, subjective time is also referred to as fluent time or internal time, the time of our lived experiences. And geometric time is objective time or external time. The interpretation of these two temporal points of view can be surprisingly rich, even controversial.[57]

Chronos and kairos. In his discussion of political and cultural analysts' fondness for identifying special "standstill" moments, Sam Anderson discusses ancient Greeks' two different notions of time— chronos and kairos. Chronos is chronological time, geometric time. Kairos is a time of significance, a defining moment, subjective time.

No matter what vision of nation-changing moments of kairos a president may have, says Anderson, most of governing is dominated by "clock-eating compromise and negotiation . . . the grinding, unglamorous time" of chronos.[58]

The future isn't what it used to be. The economist Paul Krugman reviewed Robert J. Gordon's book *The Rise and Fall of American Growth: The U.S. Standard of Living Since the Civil War*, which suggests that the great inventions accelerating prosperity gains in the decades following the American Civil War—electricity, sanitation, chemistry, engines, telephones—created a century of growth in the nation's standard of living that may not be replicated by more recent technological advancements such as television and the internet. Krugman is not 100% convinced of the argument—he characterizes his opinion as a "definite maybe"—but he does note that Gordon's analysis contrasts with recent decades' wildly optimistic visions of utopian futures enabled by technological advancements. Krugman concludes his review with the statement: "Perhaps the future isn't what it used to be."[59]

With just a single sense of time—without both geometric time and subjective time; without both chronos and kairos—the future not being what it used to be is nonsense: The future is in the future. The future will be, in the future. The future cannot be anything in the present. And certainly the future did not use to be anything in the past.

What is true is that the geometric future is at a later time than the geometric now. All of the futures— as Gordon discusses them, and as social commentators discuss them—are at a later geometric time than the geometric moment, the now, when these commentators are putting forth their ideas about possible futures.

But, even though the future takes place in the geometric future, we can still talk now about the future. We can still have a subjective conversation about later times. And in the past, we have talked about the future. And in the future we can talk about the future. Or the past.

This is completely analogous to our cognitive processes—in the past, in the present, in the future—which create who we are. The subjective time created by our sequence of cognitive quantum jumps is the time of our conscious selves, regardless of whether it was yesterday that we thought about the future, or tomorrow that we will think about the past.

Time Stands Still

"Time seemed to stop," writes Peggy Orenstein, describing the thought that her breast cancer had recurred when she felt a hard knot when her finger accidentally grazed over her lumpectomy scar.[60]

Here Orenstein's time that stopped is subjective time, not geometric time. She is not saying that, literally, the clock stopped momentarily and resumed after a period of time. She is not saying that there was a period of time which ordinarily would have registered as some positive fraction of a second, but in this case registered no elapsed time. She is not saying that, when the M^4_+ spacetime event occurred, it omitted a fraction of a second that would normally have been there.

She was saying that, in her subjective experience of time—in her own unique sequence of spacetime events—this experience proceeded as if a moment of time was omitted. In Orenstein's subjective experience of time, the shock triggered by the possible recurrence of cancer caused her to miss a beat, to have a blank spot, an empty placeholder, in what normally would have been a fully felt moment of time.

Others have attributed this sensation of time slowing down or standing still to a natural stress-induced speeding up of the brain's internal processing, triggered by hormones as an adaptive mechanism in life-or-death situations. Some athletes seem to be able to train themselves to consciously trigger this effect to heighten their ability to react to fast-moving events.[61]

What This Chapter Says About Digital Mind Math

Our natural bits of cognition—our minds' thoughts and concepts—are snippets of life, time-extended memories of three-dimensional objects moving from a beginning to an end. We experience these M^4_+ snippets, then store them in our minds with labels based on how they are multiply linked to the many other linked M^4_+ snippets already stored and linked in our minds.

The first step, possibility generation, of the three-step core cognitive quantum process—possibility generation, selection, experience—involves consideration of both the future and the past. Perhaps we may regret a past situation, and therefore—in our quest to maximize negentropy and increase information content—we speculate on how the past action could have proceeded differently, what we could have said or done that would have made the past situation more satisfying. And certainly we generate possibilities about the future, rehearsing in our mind what we will do or say.

We are adept at revisiting M^4_+ snippets that took place in the past, perhaps surgically substituting alternative hypothetical actions for what actually took place. And we are adept at extending these M^4_+ snippets into the future, hypothesizing variations on what might transpire.

Our thoughts are natural four dimensional M^4_+ 4-spheres, extending backward and forward in time. M^4_+ 4-spheres are our basic cognitive units, which we aggregate and disaggregate, dissect and amend, evaluate and experience, as our core cognitive process.

In the next chapter, we'll examine a couple of quirky cognitive experiences.

20. Déjà Vu

THE MECHANICS OF DIGITAL MIND MATH

THE ORGANIZATION OF THE MIND
- Grouping details to form concepts
  - Linking concepts

- **$M^4_+ \times CP_2$ structure: Spacetime 4-spheres, intricately linked**
- **P-adic structure: Sequences of numbered 4-spheres and their connections**

THE CORE COGNITIVE QUANTUM PROCESS
- Possibility generation: Generation of possible next thoughts

---

- **Selection: Selection of the next thought**
  - **Guided by the maximization of information content**
  - **Negentropy maximization is the only value**
- **Experience: Experiencing the selected thought**

IN THIS CHAPTER:
Sometimes our mind doesn't seem to work quite right. The experience of déjà vu—the experience that we've already seen something, or previously been somewhere—is one example. This is a mistake, isn't it? A malfunction of the mind or the brain?

We'll look at the déjà vu phenomenon—as well as another possible malfunction, the experience of going into a room but forgetting why we're there—to see what it can tell us about malfunction, adaptive behavior, and negentropy maximization.

Déjà vu is a common experience. Seemingly out of nowhere, we have a sensation in our mind that we've been here before, or heard or read this before, or seen this person, or felt this feeling.

Is déjà vu a pure malfunction of the mind or brain? Or does it have some kind of adaptive purpose?

Clearly, according to Digital Mind Math, we are highly inclined toward understanding déjà vus as a natural process of the mind, a healthy process, an experience that was selected because it tends to maximize information content, maximize negentropy.

Let's see how déjà vus can be understood this way, and let's also ask ourselves how, if at all, cognitive dysfunction fits into the Digital Mind Math scheme.

Déjà Vu, for Better or Worse

Before we get to the question of why we experience a déjà vu, let's first address what for Digital Mind Math is a much easier question: how we experience déjà vu.

With Digital Mind Math's p-adic structure, the question of how we experience déjà vu is almost trivially easy. Clearly, déjà vu results when a current experience has a cognitive informational structure that is nearly or entirely identical to a structure already within the organization of our mind.

Remember that p-adic mathematics is highly flexible about levels of detail: With p-adic mathematics, in its fully realized complete and closed informational space Ω, exponents for the p-adic power series expansion can be any rational number. This means that levels of detail can come and go—be attended to, or be ignored—as the mind generates possibilities and selects next thoughts.

So a possible next thought about a person sitting on a couch can resemble a preexisting thought about a seated person, even if the resemblance needs to ignore that the eyes are a little different, or the couch is an armchair.

The resemblance might be quite strong just because of the exact seated posture, or the lighting in the room, or the specific anxiety we're experiencing now and experienced then.

Similar to the "p-adic surgery" within our dreams that we discussed earlier, we can surgically excise certain aspects of the current experience and the preexisting memory, and find the resemblance that triggers the experience of déjà vu.

But why do we do this? Does it serve an adaptive purpose? Does information maximization result from a déjà vu?

Certainly one information-maximizing possibility is that, through the déjà vu, we're working something out: We're reminding ourselves of a similar previous experience so that we can remember how we resolved this situation in the past. Or what positive outcome followed, and how we got there. Or figure out how we got ourselves into this pickle. Or perhaps we're simply haunted by a prior memory, which our mind is always on the lookout to relive in an effort to finally come to terms with this memory, to resolve it, to work it through.

So there are many ways in which a déjà vu is an adaptive, information-maximizing experience. But is this always necessarily true? Is a déjà vu necessarily an information-maximizing experience? Could it be a mental misfire, a mistake?

Here the Digital Mind Math perspective weighs heavily in favor of the core cognitive process leading to an adaptive outcome, an outcome that maximizes information content and minimizes disorder.

Digital Mind Math does not rule out dysfunction. However. Digital Mind Math limits dysfunctional cognition to unusual circumstances, to the exception, not the rule.

Certain physical damage—brain lesions, or trauma—can result in dysfunction. As can debilitating mental illness. But the Digital Mind Math perspective weighs in heavily on the side of the negentropy-seeking mind, even when confronted with physical, mental, or emotional challenge.

The Digital Mind Math mind is resilient. It adapts to adverse circumstance. In fact, one could say it's at its best—better than most physically or consciously controlled alternatives—in the face of challenge.

Why Did I Come Into This Room?

Everyone has entered a room and forgotten why. This seems to be a phenomenon that occurs more frequently with age, but it's not clear why. How does Digital Mind Math understand this phenomenon, both in terms of how it takes place and why it takes place?

This phenomenon of forgetting is not limited to when we enter a new room, but it seems to happen most often then. Our mind seems to identify the beginning of a new concept as we cross the threshold into a new room: The physical boundary created by the doorway seems to be easily converted to a conceptual boundary also, the start of a new concept.

Before entering the new room, we had a plan. We knew we needed something, and set ourselves on the way to get it. This was all thought through quite thoroughly, and we were quite satisfied with the plan. We had a bounded concept—finished the plan—and set ourselves on the way to execute the plan. Through the doorway, into the next room, and that's when the trouble began.

Proceeding to the next room was a process in itself, a new process, that took our mind off the plan. Apparently the plan was not firmly ingrained. It was, after all, a newly developed plan. Maybe it just hung by a thread, not firmly established, not repetitively reinforced. So it was easily dislodged, disrupted by the change in focus to a new concept.

Personally, I've had some success with a couple of different strategies to resolve this predicament. One strategy is just to wait. We do know that we had a good reason for getting whatever it was that brought us into this room, and we do know that it had been as part of a well thought-out plan that this room was the place to get it. Just wait a minute. Shut off the mind. Let the mind's natural information-maximizing tendencies find the solution for us. There was good reason for why we're here, and it's only natural for our mind to revive the whole plan.

Another strategy is to have faith that there was good reason for us getting whatever we were on our way to get. Go back to our life and resume our activities. Chances are that we will very soon find ourselves back on the path that led to the unmet need. We will revive the memory because it was a memory that made sense, that emerged as an information-maximizing next experience. We haven't yet gotten what we previously decided to get, so we should very soon again decide that we need it.

At that point, we may even experience a feeling of déjà vu, realizing that we had been down this road before, and that it looked then just like it looks now.

<u>Particularly when it's a repeated event</u>. One situation in which we seem to be particularly susceptible to an experience of forgetfulness—forgetting why we entered a room, forgetting whether we've already completed a chore or an action—seems to be when the experience is one that we do repeatedly or often. Did we shut the stove off? Close the car's windows? Brush our teeth? Eat a banana?

Here the mental misfire seems to relate to the preexisting M^4_+ mental image resident in our mind due to the previous experiences of this very event. We seem to be susceptible to mistaking historical personal events for more recent personal events. Our mind finds the exact M^4_+ mental image, imbedded, perhaps prominently, in the informational structure of our mind. But the date and time stamp connecting—through a CP_2link— to the event seems more weakly stored than the event itself. This phenomenon seems particularly pronounced when the habitual event has been repeated very often in the past. And we even at times seem to mistake just the thought of the event—the mental plan to perform the event, the experience of this selected thought about the future—for the actual performance of the event.

Now these experiences are far short of conditions—hallucinations, delusions, schizophrenia—requiring psychiatric intervention. These are experiences of daily living. From the Digital Mind Math point of view, they are to be understood as deriving from perfectly accurate M^4_+ stored memories being mistakenly deployed within the mind's generation of possible next thoughts—mistaken because the mind has carelessly brought forward the M^4_+ memory without its important CP_2-linked time and date stamp, and at times even without its important CP_2 link to the label "just thought about; not yet experienced."

Once again, we are faced with the question of dysfunction. Is this mental misfire an unequivocal mental error? With no adaptive function? Does this tell us something about the downfall of the species?

Maybe.

To think otherwise, we could hypothesize some adaptive purposes for these mental misfires. Perhaps this is part of the mind's cooling system: the mind finds unimportant circumstances when it can take a bit of a break from its intense and rigorous workload, relax a bit, let off a little steam. After all, the human brain has nothing to apologize for in terms of its cooling needs compared to large-scale computers, furnaces of machinery that require massive cooling facilities. Maybe these little mind burps release a little occasional steam as a way of preventing the build-up of the volcanic forces of delusion or other severe psychiatric conditions.

Or maybe—noting that there may be some correlation between mind blunders and advancing age—these aren't really blunders at all, but rather are an adaptive refocusing of what's important in life, away from the daily grind, toward a more spiritual plane.

Just keep in mind: We are not here—within the book *Digital Mind Math*—to diagnose psychiatric conditions, to assess the state of human evolution, or to advise as to geriatric care. We are here for a much more mundane purpose: to practice using the tools of Digital Mind Math.

Informational structure and its weightings. This subject will be discussed at more length ahead in Chapter 29. But it's worth previewing now because you may have noticed that it's a needed discussion.

When we talk about some M^4_+ mental events being stored more prominently or more weakly than others, or about memories being more or less firmly ingrained or established, what exactly—within the Digital Mind Math model—do we mean by this? Where in all of the Digital Mind Math—the $M^4_+ \times CP_2$ structure of intricately linked spacetime 4-spheres; the p-adic sequences and their connections—do we find these informational weightings that permit us to describe some M^4_+ mental events as prominently stored and others as weakly stored?

The answer is: We don't. We do not find the informational weightings within Digital Mind Math's mathematics. Instead, we find them in the individual human being's brain.

The mind needs a brain, and this is one of the reasons why. Different people have different weightings of the importance of a memory or a thought, and these weightings derive from the entire past cognitive history of the individual. Which particular mental connections we formed, and how intensive these connections are, is what makes us who we are. The informational weightings are formed over the years and differ from person to person based on experience.

We'll discuss this in more depth later on, but nevertheless it was worth previewing now that it's an issue that needs to be discussed.

What This Chapter Says About Digital Mind Math

The concepts in our mind—those M^4_+ spacetime 4-spheres—have a life of their own, a structural reality imbedded within the organization of our mind. They have a prominent reality, a holistic reality, imbedded within an intricately linked informational structure.

A current situation—being constructed through assimilation and accommodation, grouping details and linking concepts—can tunnel its way through CP_2 links to another whole experience as a déjà vu.

And a jump to a new concept—through the doorway to the next room—can temporarily dislodge the details of the prior concept, until we give our minds the chance to regain its equilibrium and recover the details of the information-maximizing prior thought.

It's a good thing that the mind has evolved with a p-adic organizational structure, so that all this is made very easy.

In the next chapter, we'll continue to gain experience with Digital Mind Math's core cognitive process, and with how p-adic mathematics allows rich access to Digital Mind Math's $M^4_+ \times CP_2$ eight-dimensional mind space — not only the spacetime 4-spheres M^4_+, but also the other four dimensions, the CP_2 space of intricate linking.

21. Mrs. Wyatt Earp

THE MECHANICS OF DIGITAL MIND MATH

THE ORGANIZATION OF THE MIND

> - **Grouping details to form concepts**
>   - **Linking concepts**
>   - **$M^4_+ \times CP_2$ structure**
>   - **P-adic structure**

THE CORE COGNITIVE QUANTUM PROCESS

> - **Possibility generation**
>   - **Selection**
>   - **Experience**
>
> **IN THIS CHAPTER:**
> In this chapter we look at a few simple phrases and sentences to see how our mind organizes facts and aggregates and connects them. We examine how we group details to form concepts, how we link concepts, and how we place thoughts within the context of other thoughts.
>
> This simple example illustrates the basic Digital Mind Math organization, as well as the core Digital Mind Math cognitive process.

Here's a little snippet that strikes me as one that can be understood by organizing it in a Digital Mind Math way. It offers possibilities for diagramming how our mind is organized, as well as for identifying the steps we go through in order to organize our mind.

These four sentences are taken from a letter to *The New York Times Book Review*, commenting on an earlier review of Larry McMurtry's 2014 novel *The Last Kind Words Saloon*, based on the life of Wyatt Earp, the multifaceted character of the American West:

> "Earp took up with Josephine Marcus, the lover of Sheriff John Behan. . . [T]hey were subsequently together for 47 years. Earp, however, did have a common-law wife when he came to Tombstone . . . Her name was Mattie Blaylock, and she appears in any number of books about Earp, including 'The Last Gunfight,' by Jeff Guinn (2011), as well as in the film 'Wyatt Earp' (1994), where she is played by Mare Winningham opposite Kevin Costner's Earp."[62]

Let's take a look at what we learn from this excerpt from a letter, written by Robert S. Pingree of Concord, New Hampshire. And let's also take a look at how we learn it.

Core Cognitive Process Culminates in Reading the Letter

Wyatt Earp was already known to us, before we read Mr. Pingree's letter. Something about the letter intrigues us enough to read it seriously in order to add a little knowledge to our life. In other words, we set ourselves on the path of our next thoughts increasing our information content by adding a little to the information we already know about Wyatt Earp. That is, we are following the path of increasing information content, in this case information about Wyatt Earp.

There are many possible next thoughts that we could have after beginning to read a letter to the editor about Wyatt Earp in a newspaper book review section: We could decide, for example, that this letter is actually not as interesting as we anticipated, and therefore we move on to the next letter and to the possibility of increasing information content elsewhere. Or we could decide that letters to the editor about previous book reviews aren't for us, and move on to the current week's book reviews themselves, rather than to letters about previous book reviews. Or we could decide that the book review section isn't for us, and move on to the sports section or the business section. Or we could put down the newspaper and turn on the television, or engage in a meaningful conversation with our spouse.

But the thought that we actually decide to experience is to continue reading this letter to the editor, which we expect will give us more information than we previously knew about Wyatt Earp. In particular, we are expecting to increase our information content by increasing p, adding a bit or two of information to the information that we already have about Wyatt Earp.

Why we made the decision to read this letter and absorb its contents is because, among the decisions available to us, we judged that this route is the most appealing in advancing our information content at this moment.

This decision is made in the context of our previous history of decisions, of cognitive quantum jumps, as they have formed our informational bundles and strengthened connections among our bits of stored information.

Our previous history of cognitive quantum jumps has created a mapping in our mind of everything we've ever thought or done. This is a natural process that our mind accomplishes using its natural mathematics, p-adic mathematics.

The mind mapping permits modeling of all sorts of variations for how our mind will proceed. Each possibility—each possibility for how the mind will proceed—maps precisely to neural pathways.

The brain tests the possibilities for information maximization (maximizing order; maximizing negative entropy; creating the largest possible negative number as the measure of entropy, of disorder). Entropy is a well-defined physical process—a process of physics—and negentropy maximization is straightforward in p-adic mathematics, because p-adic size measures information content.

Anecdotes and analyses presented in other chapters of Part Three illustrate how p-adic size measures information content. For now, we note that, for the brain to analyze and select the next, information-maximizing thought, the brain's process—relying on our whole history of previously selected and experienced thoughts— determines an information-maximizing pathway.

Our past history has established connections, and the strengths of the connections, based on what we've thought before, based on our prior history of cognitive quantum jumps, based on the thoughts that were selected and experienced. This is synaptic plasticity.

We commit to taking a few moments to read this letter and absorb its contents. This is the experience of the selected cognitive path. The path has been selected based on its likelihood of increasing information content.

This experience changes our cognitive mapping, and we are now ready, with a revised, informationally enhanced cognitive mapping—both a mind mapping and a brain mapping—to begin again, to go on to our next experience.

As you can see, it is quite enticing to think of this p-adic mapping as the mapping of not only the mind but also the brain: This mapping aggregates thoughts into larger concepts, and is a mapping used both by the mind and by the brain. In other words, it is enticing to consider the possibility that the Digital Mind Math mind is a brain-ready mind, a model of the mind that maps directly to the brain. But let's set this thought aside for the moment, and continue to see how we learn more about Wyatt Earp, focusing specifically on the mind.

Next Core Cognitive Process Culminates in Increasing Information Content About Wyatt Earp

We now generate possibilities for our next thought, the thought that will follow the thought that led us to proceed to read this letter to the editor.

It becomes apparent that we need to briefly focus on what we know about Wyatt Earp, so that we can better put into context the new information that this letter will provide.

Here we see how sometimes our next thoughts are pretty small potatoes, are thoughts that aren't grand, are thoughts that seem small-picture, not big-picture, just process-oriented, when compared to the truly important thought that is our goal: adding information about Wyatt Earp's relationship status to what we already know about Wyatt.

We have our mind set on a p-adically larger thought, the thought that will increase our information content by adding information to what we already know about Wyatt Earp, specifically information about Wyatt's spouse, information that is entirely new to us.

Our mind is set on jumping from the letter to the resulting state of increased knowledge about Wyatt Earp. But we see that this cannot be done in one single jump. Our mind assesses—using its natural p-adic mathematics to model many possible future paths and to calculate each path's p-adic norm (size)—that what we need right now, before trying to take advantage of a new thought about Mrs. Wyatt Earp, is to get a good grip on our already-existing Wyatt Earp cognitive conceptualization in order to better prepare ourselves for the new Mrs. Earp information. We will be best served by strengthening in the forefront of our mind the facts we already know about Mr. Earp.

So we search through our mental database in order to focus in on our concept of Wyatt Earp—perhaps following a path from people, living or dead, to historic personages, to figures of the Wild West, to Wyatt himself.

We focus on Wyatt to see what we recollect: lawman; gambler; Wild West entrepreneur; Tombstone, Arizona; gunfight at the O.K. Corral.

In terms of Digital Mind Math, let's label the search from historic personages to figures of the Wild West as 0.354. Remember that a p-adic number such as 0.354 represents an ordered path, starting at a p-adically smaller concept or thought or entity that we label number 3, then proceeding to a p-adically larger concept that is linked to concept 3 and is labeled concept 5, then ending at the largest concept, concept 4. Concepts are p-adically larger when there are either more details (greater p), or more levels of detail (greater negative-n). Here our thought path of increasing information content proceeds by adding levels of detail.

In this particular case, this means that we first narrowed down the thought triggered by our decision to think about Wyatt Earp to mind path 0.3—people, living or dead. Then followed that path to a next level of detail 0.35—people, living or dead, who are personages of historic significance. Then continued on the path to 0.354, figures of the Wild West.

Note the use of the term "narrowed down" to explain the process of sharpening our focus from people to historic personages to figures of the Wild West. We're accustomed to thinking of this as moving from a large group of people to a smaller subset, then to a still smaller subset. This is because we're accustomed to real numbers, for which the real norm (size) is bigger for a large number of people than for a small number of people.

In p-adic mathematics, however, what is small in a real sense is large in a p-adic sense. P-adic numbers measure information content, and a thought about the more specific group of people— figures of the Wild West—has a larger p-adic size than a thought about a more general group of people. Even though in a real sense there are more people in general than there are figures of the Wild West, p-adically there is greater information content associated with the narrower, more specific group, figures of the Wild West.

We now experience this next thought, 0.354, selected based on some inner conclusion that this thought will tend to maximize information content.

There's clearly an information-maximizing element to this thought, which proceeds from a concept that in a real sense is a larger concept, a broader concept—people, living or dead—to a subset, historic personages, of that concept, and then to a subset of that subset, figures of the Wild West. Although, in a real sense, this moves us in the direction of smaller quantity—there are fewer figures of the Wild West than there are historic personages, and fewer still than there are people, living or dead—we know that what's smaller in a real sense is larger in a p-adic sense: The p-adic number 0.354 is larger than the p-adic number 0.3, so moving the focus of our thinking to figures of the Wild West has created more information content—a p-adically larger thought.

We now momentarily experience this thought—figures of the Wild West—and then will quickly move to generating next possible thoughts.

Experiencing this thought—figures of the Wild West—is an efficient way for us to proceed to prepare ourselves to receive new information about Wyatt Earp. We remember that he's a figure of the Wild West, and our algorithm for maximizing information tells us that a good next thought is to think about figures of the Wild West.

Two comments about settling on a thought about figures of the Wild West, rather than a thought about Wyatt Earp: First, our information-maximizing algorithm is constrained by our mental abilities and prior knowledge. And second, we will have the opportunity for the next thought following to focus on Wyatt Earp himself, rather than on the more general concept of figures of the Wild West.

Let's discuss these two comments, starting with why our next thought was about figures of the Wild West in general, rather than about Wyatt Earp in particular.

Our knowledge of Wyatt is shaky and sketchy. We're a little unclear about who he is and what his claim to fame is. This is based on what we know, what our previously established knowledge is, what our mind's history of mental quantum jumps has been.

We were never that into reading stories of the Wild West. We never had a Wyatt Earp cowboy hat and sheriff badge. Our doctoral thesis did not deconstruct the truth and mythology of the Earp brothers.

So our selection of our next thought as 0.354, the path to the concept of figures of the Wild West, is guided by Piaget's concept that the optimal next thought is one of moderate novelty, and by Vygotsky's concept of the zone of proximal development.

Other people—with other past cognitive histories—would at this point have different next thoughts. But this is our next thought, our personal zone of proximal development. So we're going to proceed according to our personal sense of moderate novelty. (By the way, in Part Four of *Digital Mind Math*, we'll be making some pretty wild speculations about how exactly the mind calculates moderate novelty, how the mind calculates its zone of proximal development. But for now—Part Three—it's back to Wyatt Earp.)

And now that 0.354 was the right thought for us, we begin again to generate possibilities for the thought after that.

We could jump right into proceeding to read the letter. But, to maximize information content, it would be better to have a firmer base of knowledge about Wyatt Earp.

We could go to Wikipedia and find out lots of information about Wyatt Earp. But that seems to be overkill. That seems to take inadequate account of our general knowledge about Wyatt Earp. That doesn't properly assess the likelihood that our preexisting state of knowledge is adequate for this purpose. It would actually be too novel—looking up from scratch the basic biography of Wyatt Earp— rather than the optimally moderate novelty immediately accessible right there in our mind. Going to Wikipedia for a more complete biography would—for us, for our particular circumstance, deriving from our own personal history of prior cognitive quantum jumps—be too micro a step, too small a step, not a step as large as our zone of proximal development.

To be satisfied, we need to focus on Wyatt Earp—a small and very specific subset of figures of the Wild West—which means he'll represent even higher information content, represent a p-adically larger thought: 0.3542. And our optimally information-maximizing approach will be to tap into our existing memory bank to get us just a little more comfortable about who Wyatt Earp is, before we proceed to read a little about his wives or life partners.

When we read the letter about details of Wyatt's relationships, we want to be thinking about Wyatt. We want to have our focus be on Wyatt and the top-level details that we know about Wyatt. This way, when we read the letter that will give us additional details about Wyatt, we will be increasing information content in a different way than our prior thought increased information content by going from 0.3 to 0.35 to 0.354 to 0.3542.

Going from 0.3 to 0.35 to 0.354 to 0.3542 increased information content by increasing negative-n. We increased negative-n from 1 (one digit to the right of the decimal point) to 4. Since the size (norm) of a p-adic number—that is, the magnitude of information in a thought—is p^{-n}, thought 0.3 has information content of size p (that is p^{1}), whereas 0.3542 has the larger size p^{4}.

It was this analysis, that we could increase information content to p^{4}, that led us to think in particular about Wyatt Earp.

But it is not by continuing to increase the levels of detail that we're planning to use Mr. Pingree's letter to the editor to increase our information content. We're not going to use Mr. Pingree's letter to narrow our focus, thereby increasing the level of detail and therefore increasing negative-n.

Instead, we're planning to read Mr. Pingree's letter in order to increase information content by adding to our information about Wyatt Earp. We're planning to begin reading Mr. Pingree's letter with our mind focused on Wyatt Earp—thought 0.3542—which contains a certain amount of information that we already know about Wyatt Earp. Then we're going to add another bit or two of information about Wyatt Earp, namely his marital or nonmarital status.

This will increase information content—that is increase the value of p^{-n}—not by increasing negative-n, but rather by increasing p.

We will increase information content not by delving into a deeper level of detail (higher negative-n), but rather by increasing p, by increasing the knowledge we have about the very fact—Wyatt Earp—that we're already thinking about.

Information content, which is measured by the quantity p^{-n}, increases either by increasing negative-n or by increasing p. By beginning to read the letter, focusing on Wyatt Earp, about whom we know p facts, we plan to increase what we know about Wyatt Earp to p+1 facts, or maybe p+2 or p+3.

So p^{-n} will increase, even if negative-n doesn't increase, just because p increases.

Our plan for how our thoughts will proceed is now in place: First, we will remind ourselves of what we already know about Wyatt Earp. Then, still focusing on Wyatt Earp, we will read Mr. Pingree's letter in order to increase the number of facts that we know about Wyatt Earp.

Review: Some real and p-adic clarity. This whole 0.3542 p-adic shorthand has a very real corresponding representation. Let's take a moment to ground ourselves with a description of this real representation that corresponds—adelically, deriving from both real and p-adic mathematics—to the 0.3542 p-adic representation that we have described at length above.

We have p-adically labeled as 0.3 the M^4_+ spacetime sheet for people, living or dead, which is a spacetime sheet that can be broken down into p elements or components.

Now before we go on, we need a brief reminder that we're going to act like real mathematicians and not worry about proving details that we know to be true. In particular, here—acting as the advanced p-adic mathematicians that we are—we know that big Witt vectors (also known as universal Witt vectors) can be treated just as though they are p-typical Witt vectors. So, as the world-renown p-adic experts that we are, we can think big-picture and be very loose about whether every digit in our p-adic number has the same number p of possible values (p-typical Witt vector), vs. whether there are different numbers of possible values—different values of p—for each p-adic digit (big or universal Witt vectors). When we loosely talk about the spacetime sheet levels each having p elements, we know that it's a mathematical triviality that each level might have a different value of p. After all, we're world-renown mathematicians, and we have graduate students to document in detail the p-typical-to-universal-p connection.

So the M^4_+ spacetime sheet 0.3, which is our mind's People, Living or Dead thought bubble, has p elements. As our next thought, we focus on the Personages of Historic Significance element—p-adically labeled as element 5 of spacetime sheet 0.3, the People, Living or Dead thought bubble—and we traverse through People, Living or Dead's Personages of Historic Significance CP_2 wormhole to open up to the entire M^4_+ Personages of Historic Significance thought bubble, so that our next thought is about personages of historical significance.

Now a brief reminder here that the digit assignment plan—for example, assigning 5 to personages of historic significance—is just our own personal labeling of convenience, an arbitrary digit assignment, without any indication or ordering or size. P-adic mathematics is ultrametric, and anyway our mind has evolved to use Teichmüller elements—(p-1)th roots of p, rather than whole numbers. This arbitrary digit assignment map should not be too difficult to accept, since contemporary computer science has evolved as far as it has with extensive use of arbitrary mappings of meaning to the binary digits 0 and 1.

At the p-adically labeled 0.35 thought bubble Personages of Historic Significance, we focus on one of its p elements—element 4, Figures of the Wild West—and our mind traverses through Personages of Historic Significance's 0.354 CP_2 wormhole to open up to the entire M^4_+ thought bubble Figures of the Wild West, a thought bubble with p elements.

We now focus on Figures of the Wild West's element 2—Wyatt Earp—and pass through Figures of the Wild West's Wyatt Earp wormhole to open up to the entire M^4_+ Wyatt Earp thought bubble, which has p elements. Our plan is to remain focused on Wyatt Earp—the thought bubble that is p-adically labeled 0.3542—and increase information content by increasing p.

Aside: Why is this increasing p, not increasing negative-n? This is a good question, and worthy of a side comment.

It has been entirely our choice to read Mr. Pingree's letter with the intent of increasing our preexisting extent of information regarding Wyatt Earp. We could, alternatively, have set out to increase our information content by finding out about Mrs. Earp—her childhood, her parents, her schooling, her farm chores. But that's not our intent.

Our intent is to increase information content by adding more detail to what we know about Wyatt. P-adically, this has the effect of increasing p, increasing the number of pieces of information contained in the concept Wyatt Earp, concept number 0.3542.

It would have been perfectly legitimate for us to have set out on a different quest, the quest to find out about Wyatt Earp's wife as a person, to open up a new file drawer 0.35426, a file drawer for Mrs. Earp, which we could fill with facts about Mrs. Earp, who we've become interested in after some earlier thinking about Wyatt.

Doing that would have increased our information content by increasing negative-n. It's just not the route we're taking. We're taking the route of increasing information content by increasing p (adding more points of information to what's in file drawer 0.3542), not by creating a new level of detail as file drawer 0.35426.

By the way, as an aside to this aside, I want it to be clear that Digital Mind Math is not a sexistly Neanderthal system in which wives are forever subservient file drawers to their husbands. Quite the contrary, because Digital Mind math's CP_2 space intricately links in four-dimensional space, M^4_+ spacetime 4-spheres, including Mrs. Earp, are labeled by multiple p-adic numbers, representing multiple informational paths. So Mrs. Earp is not just a detail of Wyatt's life, she is also a businesswoman and figure of the Wild West in her own right within other informational paths, each with its own p-adic label, and each leading to the same M^4_+ thought bubble Mrs. Wyatt Earp.

It's just that we've set out to learn more about Wyatt.

Reminding ourselves about what we know about Wyatt Earp. Our search of our mind has located thought bubble 0.3542 Wyatt Earp. We examine this thought bubble—open file drawer number 0.3542—and find that therein lie five facts:

- rascal

- gambler
- entrepreneur
- Tombstone lawman
- fought at the O.K. Corral

Our plan is to increase information content by increasing the number of facts that we know about Wyatt Earp. We plan to read Mr. Pingree's letter in order to add facts to file drawer 0.3542. Doing this will increase p, which represents the number of choices available for each digit of a p-adic number.

In this case, our plan is to increase p from 7 to 11.

You might now be scratching your head wondering about "from 7 to 11." You have a right to wonder about this. It is confusing, and worthy of a subsection of its own to explain.

<u>An aside about increasing p</u>. We listed above 5 facts about Wyatt Earp. We're saying that this means that p—the number of details within the concept Wyatt—is p = 7. And let's say our plan is to add two more facts about Wyatt, bringing the total number of facts to 7. We're then saying that this means that Wyatt's p will be increasing from p = 7 to p = 11.

Why is this? Why, with 5 facts, does p = 7? And why, with 7 facts, does p = 11?

To answer this, let's remind ourselves of what we learned in Part Two of *Digital Mind Math*.

The file drawer for Wyatt Earp has 5 items in it. We could, if we wanted to, open up the Wyatt file drawer, and remove and examine one of the 5 items to find out more details about it. But that's not what we want to do. We just want to consider Wyatt—5-fact Wyatt. But we are curious: What size file drawer is needed to store Wyatt's 5 facts?

To answer this, we need to know what values we could assign

to the facts about Wyatt, if we chose to go a level deeper and examine these facts. This is what tells us what size file drawer is needed, what the value of p is for 5-fact Wyatt.

We learned in Part Two that, the way p-adic mathematics works, in this case p can take on the value 0 or 1 or 2 or 3 or 4 or 5. So p can take on 6 different values if you count 0 as one of the choices, because that's how p-adic mathematics works, and because—if we did set out to find out more about Wyatt's details—we'd need the placeholder 0 if there was nothing of interest at that level but there was something of interest further down (something of interest as a CP_2 connection directly from Figures of the Wild West to, say, Gamblers, or Rascals, or Entrepreneurs, without an intermediate focus on Wyatt himself).

So the way p-adic mathematics works is that p = 7 when the choices of value are 0 or 1 or 2 or 3 or 4 or 5 or 6.

Note that this gives us an extra, unused value (p = 6). So you also need to remember that p-adic mathematics uses only prime numbers as the base of its arithmetic. So we'd still be at p = 7 if all we did was add one more fact to Wyatt's file.

But we're going to add two more facts. So the possible values for the digit to the right of Wyatt Earp 0.3542 are going to now be the whole numbers 1 through 7, plus 0, meaning that p = 7 will no longer be adequate—we will now need a p = 11 size file drawer for Wyatt Earp, advancing to the next higher prime number.

There are a couple of other points from Part Two that we might mention here, although you will not need to use either of these points for this chapter about Wyatt Earp. These are just reminders, which may help solidify some earlier mathematical concepts.

First, we do not need to keep checking about what the impact is on digits to the left if we increase Wyatt Earp 0.3542 from a p = 7 concept to a p = 11 concept. The digits to the left can each have their own p, and they do not have to change. This, you will recall, is because of the magic of universal Witt vectors, which permit the digits of a p-adic number to be based on different values of p. This is what we mean by universal Witt vectors, rather than p-typical Witt vectors.

Second, you will recall that using ordinary whole numbers 1, 2, 3, and so on is a simplification we use to make the discussion of p-adic mathematics a bit more intuitive. In advanced p-adic mathematics—which permits Witt vectors—ordinary whole numbers are replaced by the mysterious Teichmüller elements. You do not need this for our discussion of Wyatt Earp. In fact, it's time to reread the excerpt from Mr. Pingree's letter to the editor so that we may increase Wyatt Earp, concept number 0.3542, from a size p = 7 concept to a size p = 11 concept.

Increasing the number of facts that we know about Wyatt Earp. Let's reread about Wyatt's love life and decide what we'll want to save in our mind in order to enhance our informational content:

> "Earp took up with Josephine Marcus, the lover of Sheriff John Behan. . . [T]hey were subsequently together for 47 years. Earp, however, did have a common-law wife when he came to Tombstone . . . Her name was Mattie Blaylock, and she appears in any number of books about Earp, including 'The Last Gunfight,' by Jeff Guinn (2011), as well as in the film 'Wyatt Earp' (1994), where she is played by Mare Winningham opposite Kevin Costner's Earp."[63]

You know what? This is all very interesting. And it will forever shape what we feel about Wyatt Earp, what the meaning to us of Wyatt Earp will henceforth be whenever Wyatt Earp is part of a future thought, or (when we're generating possible next thoughts) part of a future possible thought.

We cannot erase from our mind's informational structure,

from our past history of cognitive quantum jumps, that we have read this excerpt. But in placing this in our mind's informational structure, we will give special prominence —put right into the Wyatt Earp file drawer—just two facts: Wyatt Earp had a long-term lover who previously was the sheriff's lover, and Wyatt Earp had earlier had a common-law wife. Everything else—the names Josephine Marcus, Mattie Blaylock, John Behan; the numbers 47, 2011, 1994; the movie details—we will store as details of the details, another level or more down, not always lit up when the Wyatt Earp concept is found, a little less accessible.

We know our limits, the limits of our mental capacity, as it interacts with our need to know. And Wyatt Earp 0.3542 is now a $p = 11$ concept, containing these seven elements:

- rascal
- gambler
- entrepreneur
- Tombstone lawman
- fought at the O.K. Corral
- long-term lover of sheriff's ex
- had previous common-law wife

Who knows when the next time will be that this thought will be part of a sequence tested as the next possible thought? Or when it will actually be selected as the next thought to experience? With this thought's two newly added elements, perhaps we'll be finding this thought along a sequence involving a social history of marriage in late-nineteenth-century America, or perhaps we'll flash on it while watching a movie about complicated marital relationships, or upon hearing a piece of local gossip.

What This Chapter Says About Digital Mind Math

Let's take a break here, take a deep breath, and emphasize several important points about the structure of $M^4_+ \times CP_2$ spacetime and the structure of p-adic mathematics, as well as several important points about Digital Mind Math's core cognitive process:

- A p-adic number such as 0.3542 representing Wyatt Earp is an individual personal label, not a universal label that everyone shares.
- The label for Wyatt Earp was preexisting within our mind (and presumably, brain)—based on our prior history of quantum jumps, our entire history of the thoughts we have had and how these thoughts have structured our mind and brain—before we read Mr. Pingree's letter to *The New York Times Book Review*.
- To be fair to Digital Mind Math, we should be sure to understand and accept that there is nothing new or mysterious in the specific activity of assigning meaning to strings of p-adic numbers. After all, we have long ago ceased to be amazed by the meaningful results that emerge from all forms of computer strings of binary digits 0's and 1's. By creating consistent rules for the operation of specific strings of binary or p-adic digits, it is a perfectly familiar concept to us that complex meaning can be created.
- For the personal life path discussed above, it best maximized information content to bring in the information gleaned from the letter to the editor as two details added to the preexisting Wyatt Earp concept, with some additional information linked as details of these details. This reflects the scope of moderate novelty and the zone of proximal development for this one personal path. Others will store

the information differently. In Part Four of *Digital Mind Math*, some highly speculative thoughts will be presented about how moderate novelty or proximal development is mathematicized as a natural p-adic process of the mind and the brain.

- We add to our information content a detail such as the name, Mattie Blaylock, as a new M^4_+ spacetime sheet connected by creating a new CP_2 wormhole to the detail "previous common-law wife" that is an element of the Wyatt Earp spacetime sheet.

- In the intricately linked $M^4_+ \times CP_2$ structure of our mind, of course the expression "Mattie Blaylock" is in turn linked to a whole network of preexisting spelling and pronunciation spacetime sheets that we long ago established in our mind and brain as part of our linguistic capabilities.

- For future references, we may use only the first link—knowing that Wyatt Earp had a previous common-law wife—and not the secondary link to her name, which may not be needed for a future thought, or may be too weakly reinforced to be accessible for a future thought. It was our choice to set up our mind's informational structure this way, based on our mind's p-adic calculation of optimal information maximization, deriving from our personal history of prior cognitive quantum jumps.

- Perhaps one day we'll have a business meeting where we're introduced to Matthew Blalock, and have a feeling that we've heard this name before, perhaps generating an irrational feeling that we can't shake that Blalock may be too much of a go-it-alone Wild West type of guy for us to go into business with.

- This entire set of new and newly linked concepts and connections is what emerged as our next thought, upon reading the excerpt from Mr. Pingree's letter. This is how we organized the new information that we read so that its increase in information content could be understood and retained. This is now part of our history of quantum jumps and will change us forever.

22. The Obsessive-Compulsive Impulse

THE MECHANICS OF DIGITAL MIND MATH

THE ORGANIZATION OF THE MIND
- Grouping details to form concepts
  - o The boundary of a thought
  - o **Complete thoughts**

IN THIS CHAPTER:

We sense when a concept or a thought is incomplete. Incomplete thoughts are often not comfortably reduced to a single point that permits them to easily become part of larger thoughts. So incomplete thoughts detract from efficient organization of the mind. And they also make the Digital Mind Math possibility generation process less efficient, by adding an order of magnitude more configurations to consider than we would have to consider if the thought were complete. In other words, incomplete thoughts interfere with information maximization, and there is value in completing them.

- Linking concepts
- $M^4_+ \times CP_2$ structure
- P-adic structure

THE CORE COGNITIVE QUANTUM PROCESS
- Possibility generation
  - Selection
  - Experience

In this chapter, we explore the obsessive-compulsive impulse, the drive for order—in particular, the drive that manifests itself by a strong urge to close an open loop, to resolve an already open issue before moving on to another perhaps larger issue.

I do not mean to romanticize mental illness. An obsessive-complulsive disorder can be a debilitating mental illness that interferes with the ordinary functions of life.

Instead, I am referring to a much milder and more common impulse that many of us feel, an impulse to complete a previous issue before moving on to a next issue.

True, the obsessive-compulsive impulse—even in its mild form—can interfere with the flow of a conversation, or a group event, or even the flow of private thoughts. But in this chapter, we discuss two beneficial impacts of the obsessive-compulsive impulse—one aspect that directly increases information content, and a second aspect that indirectly increases information content by improving the efficiency of the Digital Mind Math search process.

One might note that the obsessive-compulsive impulse has survived many generations of evolution. Perhaps there is a favorable aspect to this impulse that keeps it genetically preferred.

The Direct Favorable Impact

When someone is acting in an obsessive-compulsive manner, they just don't seem to be able to let an issue drop. They are incapable of moving on to a broader topic or a different topic, because they seem to be compelled to resolve a still-open question.

If an issue has an unresolved aspect, it has conceptual baggage attached to it. The issue cannot be cleanly referred to as a single-point concept, because that single point is always connected to an open issue, to a multipoint concept involving description or qualification or clarification.

As a result, all concepts that depend on this unresolved aspect cannot have clear edges, clear boundaries, and we can never fully resolve the p-adically larger concepts that depend on the unresolved conceptual disk.

So the obsessive-compulsive impulse—to postpone moving on until an unresolved open disk is circumscribed with a clear boundary as a complete thought—is an impulse, as annoying and distracting as it can sometimes feel, that ultimately enhances the creation of p-adically larger concepts, concepts with more information content.

The Indirect Favorable Impact

Let's remind ourselves of Digital Mind Math's central, three-part process of thinking, a process identical to the core process of quantum physics, the quantum jump: Starting at the previously experienced thought, we first generate a configuration space of possible next thoughts, then guided by information maximization we select the next thought, then we experience that selected thought.

It is with respect to the first of these three parts of the Digital Mind Math thinking process that the obsessive-compulsive impulse has its favorable indirect impact, its impact making us smarter, giving us thoughts with higher information content.

When we have an urge to close an unresolved question—even if that unresolved question is of trivial importance, of lower information content than other questions that we could address, what this urge accomplishes is to turn this open question into a closed question, into a single point rather than an open disk.

By doing this, all future first parts of the Digital Mind Math process—all future generations of the set of next possible thoughts—will become more efficient, by virtue of these generation processes not having to be distracted by all the sets of possibilities that derive from or relate to the open possibilities of the unresolved question.

We do not have infinite brain capacity. Our brain capacity is very large, but it is limited. So making one recurring aspect of our thinking process more efficient has the effect of making us smarter—

leading to outcomes with higher information content.

Ambiguity

When a statement or a concept or an understanding is ambiguous, it has not been narrowed down to a single specific point. Its specifications are not well-defined by a single specific set of instructions or criteria.

But even if a concept is ambiguous, we can still make use of it. Some people are better than others at handling ambiguity, perhaps by labeling a concept as ambiguous and moving on knowing it is so labeled, which offers a certain type of closure. If what this does is require doubling (or tripling, etc.) the number of paths that need to be explored—one for each identifiable fork in the road—this could actually be more cost-effective than going through the effort to close the issue.

Sometimes ambiguity can be a particularly effective style of management for a generalist boss of technically accomplished associates: By remaining ambiguous about technical specifications, the technical associates could be empowered to find surprising and creative solutions.

Of course, another form of ambiguity—a lazier, sloppier form, in which issues simply aren't resolved—could well result in bigger problems over time, unless the issue at hand has small downside risk and is particularly complex or expensive to resolve.

All things being equal, a well-defined, nonambiguous concept is better than an ambiguous concept, but an ambiguous concept is better than no concept at all.

Ambiguity has its place. And some people are able to proceed fruitfully with ambiguous concepts, even making a virtue of it. Obsessive-compulsive traits pull in the opposite direction. It's difficult to generalize in favor of one approach or the other. Within limits, someone comfortable with ambiguity may be easier to work with, but someone driven to closure may produce more efficiently.

What This Chapter Says About Digital Mind Math

The urge to complete a thought—to establish the boundaries of a thought, so that we're comfortable referring to it as an established concept without restrictions or qualifications, without ifs, ands, or buts—is an evolutionarily favorable urge, which makes us smarter by enhancing the information content of our conceptual framework, and by making our thinking process more efficient.

In the next chapter ,we explore a different psychological pattern—that of the archetypical absent-minded professor—to see what makes him tick, to see why his absent-minded behavior is a natural consequence of Digital Mind Math.

23. The Absent-Minded Professor

THE MECHANICS OF DIGITAL MIND MATH

THE ORGANIZATION OF THE MIND
- Grouping details to form concepts
  - Linking concepts
  - $M^4_+ \times CP_2$ structure
  - P-adic structure

THE CORE COGNITIVE QUANTUM PROCESS
- Possibility generation: Generation of possible next thoughts

---

o **Adding detail**
o **Migrating to a connecting thought**

IN THIS CHAPTER:

P-adic mathematics, and therefore Digital Mind Math, quantifies the size (norm) of a p-adic number as p^{-n}. This means that there are two ways for a p-adic number to become larger: either higher p, or higher negative-n. Said another way, information content increases by adding detail (higher p), or by increasing the extent of interconnection (higher negative-n). In this chapter, we look at the caricature of the absent-minded professor—absorbed in the great detail of his area of specialty, but deficient in his ability to make common-sense connections.

- Selection
- Experience

Who is the absent-minded professor? What are the character

traits and behaviors of this archetype, this classic image of the smart man who can't get out of his own way?

First of all, the absent-minded professor is a professor, an accomplished academic expert, master of his field. He has wide-ranging command of his field of study. At his fingertips are vast quantities of historical and up-to-the-minute facts and data that fall within his field.

P-adically, we may envision the professor as operating in a high-p environment within the conceptual disk that is his field of expertise. The professor has command of a great deal of information about his field. For operation within a specific academic field, a higher number of facts results in higher information content. This is a sign of the professor's expertise and why he earned academic tenure.

But—when the the source of his information content is disproportionately the high value of p for his one particular field of specialty—the professor may well be deficient in two other important paths toward information maximization.

One of the professor's deficiencies is that he may have low-p mastery of numerous other fields or concepts. He may not have adequate working knowledge of the field of driving a car, or hammering a nail, or barbequing on a gas grill.

The professor may not be sufficiently familiar with the components of these processes. As a result: cars may crash; fences may collapse; gas grills may explode.

During his lifetime to date, the professor has disproportionately selected as his next thoughts those that relate to his specialized field. With our lives finite, this has resulted in a relative lack of mastery of practical skills. The professor is very smart—has an information content with a high p^{-n} —but accessing this high p^{-n} requires the scope of inquiry to be the specific scope of the professor's field of expertise.

The second of the professor's deficiencies may also derive

from his disproportionately large history of thoughts focused on his academic specialty, but this second deficiency relates instead to his low-negative-n tendencies. The professor is an encyclopedia of facts, but he is weak on interconnections. This may well exhibit itself as a lack of common sense.

The professor may be so focused on setting the hour and minute of his alarm clock that he forgets to turn on the alarm. Or so focused on counting the shingles on a roof that he walks backward over the edge. Or so focused on the angular momentum of a spinning propeller that . . . well, you get the picture.

Here it is not insufficient detail about an alarm clock or a roof or a propeller that results in the professor's absent-mindedness. He has the details. It is the connections that he's missing: There's another step—turning on the alarm—after the details of setting the alarm's hours and minutes. Another step while measuring roofing (being mindful of the roof's edges). Another step (keeping your body parts out of the propeller's path) when measuring angular momentum.

This low-negative-n pattern can even extend to the professor's own area of expertise. Perhaps this makes him a bad teacher—incapable of breaking information into levels of detail that can be absorbed systematically, piece by piece; incapable of conveying information in digestible chunks. Or perhaps this low-negative-n pattern will even limit the professor's creativity in his own field, so absorbed in the quantity of details that he misses potentially rich connections to related fields or even within his own specific field.

What This Chapter Says About Digital Mind Math

For a given level of information content, if there's an unusual preponderance of that content coming from high knowledge of detailed information (high p) about a particular specialty, it's only natural that the contributions to total information content from other sources will be unusually low—either low p (low level of detailed mastery) in numerous other subject areas, or a low-negative-n pattern (low interconnectedness of thought).

This is the plight of the absent-minded professor.

In the next chapter, we'll examine what could be another bad habit of our poor absent-minded professor: telling jokes. Let's see how jokes—even bad jokes—can illustrate some aspects of Digital Mind Math.

24. Jokes I—Finding the Boundary

THE MECHANICS OF DIGITAL MIND MATH

THE ORGANIZATION OF THE MIND
- Grouping details to form concepts

> - **The boundary of a thought**
> - **Complete thoughts**

- Linking concepts
- $M^4_+ \times CP_2$ structure
- P-adic structure

THE CORE COGNITIVE QUANTUM PROCESS
- Possibility generation: Generation of possible next thoughts
- Selection: Selection of the next thought

> - **Experience: Experiencing the selected thought**
>
> **IN THIS CHAPTER:**
> We look at our cognitive processes that create boundaries around the elements of a thought. And we look at our drive to stabilize bounded elements so that they can be treated as a complete thought.
>
> The completed thought is experienced as the next thought, then remains within the organization of the mind as a complete bounded concept, intricately linked, forever changing the mind's informational structure and its weightings.

Apologetic Introduction: Jejunosity

Jokes are not necessarily the highest form of discourse. In fact, some may consider jokes—in particular, the two jokes you are about to be presented with, in this chapter and the next—as particularly childish or adolescent in their appeal, juvenile, jejune.

Here one may be reminded of a scene from the the Woody Allen movie *Love and Death*, a satire of Russian historical events and Russian literature. Diane Keaton's character, Sonja, has accused Woody Allen's character, Boris, of making an "incredibly jejune" comment. Boris fights back: "You have the temerity to say that I'm talking to you out of jejunosity? . . . I'm one of the most june people in all of the Russias."[64]

Apologies in advance for the high jejunosity of these jokes. But remember, I'm sure there are some people who find them "june."

The Joke

Here's a joke for you.

Maybe you've heard it already.

If so, maybe you can play along, just to be polite.

QUESTION: What does the agnostic dyslectic insomniac do?

ANSWER: [We're going to analyze this before we get to the answer.]

How do we go about trying to figure out the answer to this riddle?

The Digital Mind Math of Jokes

To examine this joke, we first look at the question that is asked within the joke.

There is already a certain amount of humor in the question: a string of three- or four-syllable words, grown-up words indicative of a sophisticated vocabulary and a level of comfort with a rapid-fire string of big words. Yet we recognize the intention to juxtapose this opening sophistication with some closing jejunosity.

There's a pleasant musicality to the words, a rhythm, with some similarity to a poem or a song.

And then after these initial, titillating sensations, we get to a purely intellectual challenge, requiring us to think about the three main words and generate streams of denotations and connotations: What does each word mean? Is there more than one meaning? Are there common expressions in which each word is used? What habits and behaviors are linked to each word?

Each of these questions causes our mind to branch off in the directions of multiple interconnections. We think of the dictionary definition of agnostic. We wonder if there are sayings or expressions about agnostics, about famous agnostics, about how agnostics differ from athiests and from believers.

We do the same for dyslectics and insomniacs.

We put one idea—Christopher Hitchens—out into a trial thought disk to see how dyslexia and insomnia might fit in. OK, Hitchens was an antitheist, different from an agnostic, but maybe it's close enough. Hitchens. Hitchcock? Hitchcock the dyslectic? Was he dyslectic? Was he an insomniac? Hitching post? Got hitched?

Nothing's working with Hitchens. Back to square one. Can't read. Can't sleep. Toss and turn. Turn pages. Turn away from religion. Turn, turn, turn. A season. Every purpose under heaven.

Now we're getting somewhere. Who sang that? Pete Seeger? The Byrds? The birds—dyslexia—Hitchcock!

The goal. The goal seems to be to find a disk that neatly— aesthetically, minimally—encloses the three concepts—agnostic, dyslectic, insomniac.

A good answer will have no excess baggage, no loose terms. It will enclose the three concepts completely but snugly. The answer is a disk that neatly and intrinsically encloses these three concepts.

The process. The process is straight out of Piaget's assimilation and accommodation, which in turn is precisely modeled p-adically as variations on increasing p and increasing negative-n.

We've got three concepts in a thought disk, but it's not a satisfying thought disk—it's not stable, has no equilibrium, no equilibration. We add data—increase p, increase the data points in our thought disk. We temporarily branch off to a related concept, or increase negative-n by going to a greater level of detail about dyslexia, or insomnia, or agnosticism. We accommodate the structure to the possibility that two or three disparate terms—hitch, or turn—might be combined into a new disk, maybe the perfect, equilibrated disk, not too big, just big enough, explaining, but not overexplaining.

The pleasure. So why do we experience a certain joy in our attempt to figure out the answer to this riddle? And if we don't figure out the answer, and instead are told the answer, what's the source of pleasure in that?

We may here look again to Piaget, and also to Vygotsky, for our answer: The quest for equilibration, equilibrium, even if momentary, is an inherent drive of life. This is a biological drive, as well as a cognitive drive. And play—riddles, jokes, intentional play—is a natural pathway for this drive.

The Riddle's Answer—Finding the Equilibrated Bounded Concept

So what does the agnostic dyslectic insomniac do?

ANSWER: Lie in bed all night wondering if there really is a dog.

This is a wonderfully efficient answer, a marvel of conciseness and simplicity: Five simple words—lie in bed all night—capture the insomnia. Seven words—one backward—represent agnosticism: wondering if there really is a God. And just one three-letter word—dog—to cleverly represent dyslexia.

What This Chapter Says About Digital Mind Math

We are driven to increase information content, and we derive pleasure from aesthetically stated, streamlined, efficient thoughts. A completely satisfying answer to the riddle leaves us with perfect momentary equilibration. The answer has bounded three elements into a complete thought, which closes a disk to become a point of pleasantness in our informational structure, and a point of vocabulary exercise, and a point of analysis and synthesis.

We will always have this point in our informational structure, linking to other examples of pleasantness, and of vocabulary-building, and of analysis and synthesis. And—in our future generation of possible next thoughts—we will be able to efficiently reference this point as an element of next possible thoughts, thoughts that continue to increase information content.

In the next chapter, we'll examine a second joke, this one helping to illustrate another facet of maximization of information content. Hopefully, you will again forgive the jejunosity.

25. Jokes II—To a P-Adically Larger Bounded Space

THE MECHANICS OF DIGITAL MIND MATH

THE ORGANIZATION OF THE MIND
- Grouping details to form concepts

- **Linking concepts**

- $M^4_+ \times CP_2$ structure
- P-adic structure

THE CORE COGNITIVE QUANTUM PROCESS
- Possibility generation: Generation of possible next thoughts
  - Adding detail

- **Migrating to a connecting thought**

IN THIS CHAPTER
This joke is cognitively satisfying because it brings us to a p-adically larger thought space, by migrating to a component of the most recent thought.

- Selection: Selection of the next thought
- Experience: Experiencing the selected thought

Here's another joke:

Sandy gets a new car, and just to be safe, she asks a Catholic priest, a Protestant minister, and a Jewish rabbi to bless it.

The priest sprinkles the car with holy water.

The minister has us all circle the car, hold hands, and sing from a psalm.

The rabbi cuts an inch off the tailpipe.

The Digital Mind Math of Jokes

Here we start with a equilibrated concept—a priest, a minister, and a rabbi. This triumverate of religious leaders has long been a staple of American ecumenical jokes, even if in future decades more diversity—Hindu, Muslim, Buddhist, animist, atheist—will inevitably expand the scope of ecumenicalism.

The joke begins as a complete thought, but immediately, at the second sentence—the priest's holy water—we see that the balance is to be interrupted. The satisfying equilibrium that we're in continual search of—the complete triumverate of the twentieth-century American ecumenical joke—has been given to us, only to be immediately broken, immediately dislodged, on the path to an even higher level of equilibration.

This is the process of cognition,and the process of life: achieving a satisfying but temporary equilibration, only to disrupt it and start again—through assimilations and accommodations—in the hope of achieving a new higher equilibration.

P-adically, our first concept is the complete concept—the classically American religious triumverate of the priest, the minister, and the rabbi.

That's the thought we start from, the prior thought.

The next thought goes into detail about the priest. We select an element from the bounded concept priest/minister/rabbi, then we branch into a detail—a p-adically larger thought—that the priest sprinkles water.

But there is no equilibrium yet. We know what we need to

achieve equilibrium, to find a more detailed thought—a p-adically larger thought, a thought with higher information content—that branches from all three, not just one, of the components of the original thought.

We're restless, unsatisfied. The minister must be next. And there he is, having us circle the car, holding hands, singing a verse from a psalm.

This is exciting. The punchline is next. The rabbi will provide a p-adically larger detail, a detail which will complete a new thought, a closed bounded thought, at a higher level of equilibration, a higher level of information content.

What will be the characteristics of the rabbi's detail? What can the rabbi do for us that will complete the p-adically larger thought in a way that seals the deal, that signifies closure?

It is not enough for the rabbi to sprinkle holy water or lead us in communal prayer. This would be an encyclopedia entry, not a joke. We need a killer line.

The rabbi must stretch us a little, give us some moderate novelty, lead us within a zone of proximal development.

Our rabbi—the one who cuts an inch off the tailpipe—takes us to a slightly off-color zone, not just a dry encyclopedia entry (for example, the rabbi reads from the Book of Deuteronomy . . .), but nothing vulgar either, nothing that goes too far. A circumcision seems just right: It's in the Bible, but about a private body part. Perfectly moderate novelty.

The effective and correct extent of novelty depends on each individual's existing informational structure as it has been created by her entire past history of cognitive quantum jumps. And p-adic mathematics offers flexibility for how moderate the novelty will be, how proximal the development. (This point about flexibility offers an opportune moment for a tantalizing aside: In Part Four of *Digital Mind Math*, we'll be looking to a particular aspect of advanced p-adic mathematics—the converse of Hensel's Lemma—to draw a surprising conclusion about a limitation on this flexibility that provides great efficiency to the cognitive process.)

If we were dealing not with p-adic numbers, but with real and complex numbers, we have only the grossest, the clumsiest, of tools for going to a next level of detail, for advancing our exponent negative-n.

With real numbers—and their closed complete number system, the complex numbers—exponents n are integers, positive and negative whole numbers and zero: 1, 2, 3, -1, -2, -3, and so on, including 0.

This does not give us a lot of flexibility for reaching optimally moderate novelty.

If our brain—and the mind through which we sense the brain—had been set up using real and complex numbers, imagine how limited our choices would be to figure out what the rabbi did. We would have to negotiate a rigidly fixed pre-existing conceptual structure where every level of detail of all of our thoughts could be only at the same level of detail, or one whole level of detail greater or less, or two whole levels of detail greater or less, and so on.

This would not be a robust brain. Or a robust mind.

And this is an aspect of why our brains did not evolve within a real or complex mathematical structure.

This is why it was evolutionarily advantageous that our brains evolved with a p-adic structure, within the fully closed and complete p-adic numerical space omega Ω, in which levels of detail—p-adic exponents n—are not restricted to positive or negative whole numbers, but instead may be any fractional level of detail, any decimal exponent. The exponent n can be any rational number, not just any integer.

What This Chapter Says About Digital Mind Math

With thought space being organized p-adically, within the fully realized p-adic numerical space Ω, we are richly able to tweak the novelty of how the rabbi will complete our thought, trying out increments of novelty, testing and retesting until we reach a possible next thought that maximizes information content at the same time that it is complete.

In the next two chapters, we'll look at a different aspect of popular entertainment—this time mysteries, rather than the past two chapters' jokes—to continue to gain experience with Digital Mind Math's organizational structure and core cognitive process.

26. Mysteries I—Finding the Boundary

THE MECHANICS OF DIGITAL MIND MATH

THE ORGANIZATION OF THE MIND
- Grouping details to form concepts

> o **The boundary of a thought**
> o **Complete thoughts**

- Linking concepts
- $M^4_+ \times CP_2$ structure
- P-adic structure

THE CORE COGNITIVE QUANTUM PROCESS
- Possibility generation: Generation of possible next thoughts
- Selection: Selection of the next thought

> - **Experience: Experiencing the selected thought**
>
> **IN THIS CHAPTER:**
> As we did in Chapter 24 for a joke, this time for a mystery story we again look at our cognitive processes that create boundaries around the elements of a thought. And we look at our drive to stabilize bounded elements so that they can be experienced as a complete thought, enhancing the mind's informational structure into which this thought now imbeds.

For fans of mysteries, there can be great pleasure in the unfolding of the plot. The mystery fan is absorbed in the details as they are unfolded—the names, the locations, the habits, the interactions, the behavioral tics, the unusual actions and reactions.

Which of these details will figure into the overall plot? Are some of these details red herrings? Will there be a satisfyingly complete resolution, in which the author has managed to tie all these details together, with no loose ends, where all of these odd and seemingly unrelated clues actually fit together into a single coherent whole?

In a well-constructed mystery, everything should be explained at the end. There should be no red herrings, no false trails that the author has inserted arbitrarily, just to put us off-track.

Sherlock Holmes himself makes this point in the 1946 film *Dressed to Kill*: "One of the first principles in solving crime is never to disregard anything, no matter how small." [65] Sherlock lets us, the viewer, know that we, along with the great Sherlock, are to observe everything. Everything will matter. The resolution will tie up every loose end.

As *Dressed to Kill* begins, a Dartmoor Prison inmate is building music boxes in the prison shop. Is his fellow inmate's interest just idle chatter? Or is the fellow inmate a Scotland Yard plant, trying to draw out a secret? What secret?

The music boxes, plain and simple as they are, are being sold at auction. The three boxes are purchased at low prices by three different bidders. A man arrives at the auction late and inquires who purchased them. All three purchasers are sought out. Two murders ensue.

Sherlock gains possession of one music box, and with his perfect memory, after hearing the second music box just once, has memorized the tune of the second, which is now in the possesion, along with the third, of the elegant but evil Mrs. Hilda Courtney of the Park Mansions on Bryantson Square.

Sherlock, and presumably Mrs. Courtney too, have figured out that there is a code hidden behind minor differences in the three music boxes' tunes.

Both Sherlock and Hilda can partially figure out the hidden message, but they both need a third box. Hilda, but not Sherlock, knows what the purpose of the code is: Hilda and her crew are co-conspirators with the Dartmoor prisoner.

Aside from the murders (which can be incidental crimes in this genre), what is the real crime, the crime that can make double homicide incidental? We still don't know. How can Sherlock prevent the adverse outcome, whatever it is? How will Sherlock escape death? This time, will the bad guys win?

The plot pulls us along—new clues, new unanswered questions.

With his typical cleverness, Sherlock catches Mrs. Courtney in a blunder: She has left behind a partially smoked cigarette which only Sherlock is capable of analyzing—tracing the tobacco back to the shop where it was sold, leading Sherlock directly to Bryantson Square.

But wait. It was Hilda Courtney who has set the trap. She had read Sherlock's monograph on the ashes of 140 different varieties of tobacco, and lured Sherlock to her home.

A fatal error for Sherlock?

He's taken away to a slow death—poisonous gas asphyxiation while hanged from a garage ceiling with his hands tied, his mouth taped, and a car running.

Spoiler alert: Sherlock escapes! And he finds out what is motivating Mrs. Courtney and her crew: The plates for the Bank of England five-pound note had been stolen. England will be ruined if Mrs. Courtney locates where they were hastily hidden after their theft.

It looks like Mrs. Courtney beats Sherlock and Scotland Yard to the punch, but no . . . Sherlock has figured it out (with an inadvertent clue from the bumbling Dr. Watson).

England is saved!

Enclosing It All

We know, in a Sherlock Holmes movie, that clues can be anywhere. In any scene, something said, something worn, some furnishing or piece of dust—any minor factor, or combination of factors—can figure into solving the whole mystery.

In the course of the movie, we wonder: Is that the key clue? Is the mystery solved? Only to find that there are more twists.

Afterward—with the crime solved, the nation protected, the criminals in captivity—we are satisfied. The story is complete.

The urge to maximize information content is manifested by our urge to close a circle around the movie's entire plot and all of its hints and clues and implications.

We want to see the elegance of the author's creation, of the director's creation.

Does every action by every character now make sense?

Can we enclose the whole story within a sensible, integrated storyline, that lets us put this story to rest, satisfied, complete?

This is what finding and creating and certifying the boundary does for us: It allows us to place this mystery into our mind as a closed point, an efficiently stored memory, interconnected in many directions, but a point nonetheless.

Perhaps we will revive this memory when a friend tells us about a rare tobacco scent. Or when we see a London travel brochure photograph that could be Bryantson Square. Or when we look at a five-pound note. Or maybe a crisp five-dollar bill that looks like we're the first to handle in circulation.

Or maybe someone will trick us—set us up—in a way that's reminiscent of Mrs. Courtney's taking advantage of knowing Sherlock Holmes' tobacco ash monograph.

We have made a tidy investment in our lifetime store of information: A well-defined, bounded concept—the experience of the Sherlock Holmes movie *Dressed to Kill*—can be opened up and re-experienced as a rich and pleasant memory when triggered from many different directions. It can offer us insight or comfort.

And by being well-defined and bounded, this memory is an efficient resource in Digital Mind Math's continual search process. We don't need to open this memory up and examine it until and unless it's the optimal next thought that our mind selects to experience.

This is a well-tagged, well-labeled thought, interconnected in many directions to other thoughts and feelings and concepts, accessible when we need it, and until then efficiently labeled, filed, and waiting in the wings.

What This Chapter Says About Digital Mind Math

The M^4_+ spacetime 4-sphere film *Dressed to Kill* is stored in our memory, labeled with tags for all of the characters, actors, scenes, props, dialogue, clues, plot lines, and emotions that were meaningful to us as we watched this film.

P-adic mathematics is highly efficient at organizing these tags, as they are created real-time and as they are accessed upon subsequent reflection, keeping track of the characteristics of the whole and its many parts, and how these characteristics are repeated and reflected in memory segments elsewhere in our mind.

As we generate possibilities for future thoughts, anything and everything that we thought about *Dressed to Kill* will be part of the configuration space of possible next thoughts. If relevant—that is, if selected as an information-maximizing next thought—our mind will travel through the mind's real CP_2 tunnels and replay the selected *Dressed to Kill* scene as our next experience. But even if not selected to be experienced, *Dressed to Kill* will always reside within our mind's informational structure, forever influencing who we are. We'll always have *Dressed to Kill*.

The accomplished twentieth- and twenty-first-century British mystery writer P.D. James said: "In a sense, the detective story is a small celebration of reason and order in our very disorderly world."[66] In the next chapter, we'll discuss another aspect of detective stories' and mysteries' creation of order in a disorderly world.

Creation of order. Minimization of disorder (entropy). Maximization of order (negentropy).

Our mind is guided by the maximization of information content.

27. Mysteries II—To a P-Adically Larger Bounded Space

THE MECHANICS OF DIGITAL MIND MATH

THE ORGANIZATION OF THE MIND
- Grouping details to form concepts

> o **The boundary of a thought**
> o **Complete thoughts**

- Linking concepts
- $M^4_+ \times CP_2$ structure
- P-adic structure

THE CORE COGNITIVE QUANTUM PROCESS
- Possibility generation: Generation of possible next thoughts

> o **Adding detail**
>
> **IN THIS CHAPTER:**
> Similar to the joke that we examined in Chapter 25, a mystery story can be cognitively satisfying because it brings us to a p-adically larger space. Chapter 25's joke did this by migrating to a more detailed level. In contrast, this chapter's mysteries keep us at the same thought, but nevertheless bring us to a p-adically larger space, by adding a detail—
> a twist of plot.

o Migrating to a connecting thought
- Selection: Selection of the next thought
- Experience: Experiencing the selected thought

There is a second type of movie mystery, which goes beyond

the unfolding of clues, to a level of deliberate deception of the viewer.

This second type of mystery may for much of its length look just like the first type—an unfolding of clues, in which the viewer, along with the detective-role protagonist, views clues as well as red herrings, and tries to make sense of it all.

In the first type of mystery, and often for much of the second type of mystery, the viewer is in the detective role, being shown clues which—if the viewer is clever and observant enough—the viewer is right alongside the detective, solving the mystery, maybe even before the movie's detective solves it.

But in the second type of movie, there is a twist. Perhaps a twist so late in the movie that there is a surprise ending. The twist changes the viewer from simply being a partner of the detective, a surrogate detective, to being something else, an outside observer.

Plot Twists

The AMC television network (originally the American Movie Classics channel) has on its website its evaluation (written by Tim Dirks) of the greatest movie plot twists and surprise endings.[67]

In an astounding work of cinematic analysis, Dirks has summarized the plot lines of hundreds of movies with significant plot twists, highlighting the key points for each movie.

The surprise ending or plot twist is a critical aspect of the fun experienced in watching these movies, and I don't want to spoil your fun. So no "spoiler alerts" here—with one exception, I'll avoid discussing specific movies. And this one exception has such a rich and complex set of plot twists that I won't be revealing the ending, because it remains so unclear to many viewers and critics, and certainly to myself.

Let's warm up by seeing how we can think through the general nature of the appeal of a great plot twist or surprise ending. Dirks helps us here by identifying eleven major types of plot twists, with film examples of each. A narrator may be revealed toward the end of the movie as having fabricated the story that the viewer saw. Or perhaps it is a flashback to an earlier point in time that explains everything. Or a sudden revelation puts everything into place. Or it was a dream, or a conspiracy, or a hidden identity, or a character thought to be dead actually being alive, or a character thought to be alive actually being dead. [68]

What is the appeal of the end-of-movie plot twist?

When we're most of the way through a movie, and we still have a bunch of facts that are only loosely grouped into a structure, we're looking to achieve some kind of closure, to create a conception of the movie as a complete thought, a bounded concept, by finding a unifying theme or conceptualization that satisfactorily circumscribes its pieces. But the facts just aren't adding up. There just doesn't seem to be any way that a neat boundary can be circumscribed around these details to create a single complete, bounded, equilibrated thought.

When revealed, it is the plot twist, in any of its various forms, that allows this: One extra detail allows us to create an overarching consistent framework. One sharp and pointed revelation permits the reconciliation of so many previously unresolved issues that we are in awe of the extent of information maximization that we have experienced in this single next thought.

Mulholland Drive

David Lynch's 2001 film *Mulholland Drive* is a masterly combination of dreams, fantasy, flashbacks, and imagination—where four characters may actually be three, or two, or one; with murder, suicide, and mobsters thrown into the mixture. It's characterized as one of the most confusing movies of all time.[69, 70, 71, 72, 73] Its director, David Lynch, included ten clues to unlock the mysteries with the DVD of the movie,[74] and many film analysts have published essays analyzing his clues and otherwise interpreting the film.

Why on earth would anyone want to work this hard to understand a movie? What can possibly be the appeal of a movie that has hundreds of pages written to explain it?

To answer this, we can look to Digital Mind Math, including the speculations ahead in part Four as to what—mathematically, p-adically—optimizes information maximization.

Before we get to Part Four, we have already noted living systems' tendency toward order or negentropy, this tendency having multifacted manifestations, including as the mind's selection of the next thought to experience being guided by information maximization. So a movie that is confusing interests us when it provides an opportunity for us to participate in its tidy information-maximizing explication.

But what is the optimal level of confusion that makes us love its explication?

Obviously, this depends on the individual and her entire past history of thoughts, of cognitive quantum jumps. But is there any general guideline that can be provided?

Here—and we will be postponing the detailed development of this point until Part Four—we have a speculative answer provided by a bit of p-adic esoterica: the converse of Hensel's Lemma.

This is too much of a story to go into right now. But for the time being, let that thought sink in a bit: A few more chapters under your belt, and you will speculate with me as to whether there is a mathematical formula for the best confusing movie. Or the best way to converse with people. Or the best way to live your life.

What This Chapter Says About Digital Mind Math

It's been said: "I love . . . mysteries even if I've seen them before. . . Honestly, I will watch them over and over. Even if I know who did it. I still like to figure out what the foreshadowing is—the clues they are dropping."[75]

We derive enjoyment from intellectual pursuits, including the pursuit of maximizing informational content as measured by the p-adic norm applied to the structure of the mind.

Negentropy maximization guides our selection of thought and our cognitive experience.

In the next chapter, we'll look a little more closely at our experience of thoughts, once selected by our mind as information-maximizing: Who or what is it exactly that is experiencing this thought? In the quantum physics that is generally accepted at the elementary particle level, which in Topological Geometrodynamics and Digital Mind Math is also the model for the mind's core cognitive process, the collapse to a single experience results from a measurement or observation. By whom? Where is I?

28. Where Is I?

THE MECHANICS OF DIGITAL MIND MATH

THE ORGANIZATION OF THE MIND
- Grouping details to form concepts
  - Linking concepts
  - $M^4_+ \times CP_2$ structure
  - P-adic structure

THE CORE COGNITIVE QUANTUM PROCESS
- Possibility generation: Generation of possible next thoughts
  - Selection: Selection of the next thought

- **Experience: Experiencing the selected thought**

IN THIS CHAPTER

Our conscious focus is directed to the momentary experience of the selected thought. This is where "I" is, where our conscious presence is.

Our history and progression of cognitive quantum jumps—conscious experiences—is what forms our sense of self and who we are.

In generally accepted standard quantum physics, a core process at the elementary particle level involves particles not having a specific location until they are observed or measured. There are several different ways to refer to an elementary particle's realization at a specific location upon observation or measurement, one of which is the quantum jump: At the quantum jump, an elementary particle's position is reduced from a configuration space of possible positions, each with a known probabilty, to a specific position. This is very strange but nevertheless unquestionably true. All contemporary physicists agree with this. But, almost a century after Albert Einstein and others first posited this, exactly what this means and how this works is still under debate.

In Topological Geometrodynamics (TGD)—the radical theory of modern physics upon which Digital Mind Math is based—the quantum jump is a phenomenon at all scales, not just at the scale of elementary particles. TGD is a theory of cosmology and the universe, particle physics, biophysics, and cognition. As applied to cognition, standard quantum physics' quantum jump for elementary particles is also the core cognitive process for TGD and therefore for Digital Mind Math.

Our sense of "I"—who I am, what this entity is that feels like "me"—derives from the final phase of the three-phase core cognitive quantum process, namely the phase in which we experience the selected thought.

The earlier two phases of the core cognitive quantum process—possibility generation, and selection guided by maximization of information content—are biological processes, organized p-adically, and not under conscious awareness.

Consciousness—the same consciousness that in standard quantum physics triggers a particle's reduction from a configuration space of possibilities to a real specific location—is the real experience of the selected next thought.

Our experience of time—subjective time—is our sequence of cognitive quantum jumps.

Who we are is our entire history of prior cognitive quantum jumps.

"I" is located at the sequence of real experiences that result from the sequence of thoughts that are, at successive cognitive moments, selected as information-maximizing from a configuration space of possible thoughts.

What This Chapter Means About Digital Mind Math

Our process of cognition in Digital Mind Math is a process that also takes place for the elementary particles of universally accepted quantum physics, and (for radical theories of physics such as Topological Geometrodynamics) at macroscopic and macrotemporal scales also.

The organization of information that results from our sequence of quantum jumps also follows core processes of quantum physics, as discussed in the next chapter.

29. Informational Structure and Its Weightings

THE MECHANICS OF DIGITAL MIND MATH

THE ORGANIZATION OF THE MIND
- Grouping details to form concepts
  - Linking concepts

- $M^4_+ \times CP_2$ structure: Spacetime 4-spheres, intricately linked
- P-adic structure: Sequences of numbered 4-spheres and their connections

---

THE CORE COGNITIVE QUANTUM PROCESS

- **Possibility generation: Generation of possible next thoughts**
  - **Selection: Selection of the next thought**
    - **Guided by the maximization of information content**
  - **Experience: Experiencing the selected thought**

IN THIS CHAPTER:
The mind is structured as intricately interconnected ordered paths of spacetime 4-sphere events. P-adic mathematics provides the numbering system for this structure of the mind.

After experiencing a prior thought, the mind generates possible next thoughts. One of these next thoughts is selected to be experienced, guided by maximization of information content as measured by maximization of the p-adic norm.

In this chapter, we look at how interpretations of quantum physics offer guidance to address: What possible paths are available? How does the mind apply probabilities reflecting the likelihood of the

> various steps in each possible path? What is the meaning of the paths
> not taken?

In Digital Mind Math, each thought is a specific finite time-ordered sequence—from beginning to end—of three-dimensional spatial objects from moment to moment, progressing from scene to scene, which we experience as we might experience watching a video clip or interacting in the world.

As such, each thought may be experienced as the progression through time of collections of objects as they vary in three spatial dimensions. In other words, we may experience each thought according to its M^4_+ structure, its structure within physical space.

But we can't go about all day just experiencing thoughts. We don't just experience thoughts randomly. There is a method to our madness. Or to our sanity, actually.

In between experiencing thoughts (as their M^4_+ spacetime 4-spheres), we need to figure out the best next thought to have.

To do this, we take advantage of a very efficient numbering system—p-adic mathematics—which efficiently labels every possible connection among every possible thought, including all of the connections, and all of the connections to the connections, to every connecting thought.

Whether or not specific connections exist depends on our individual past cognitive history, which has created our collection of connections, and which has also created varying strengths for these connections.

The p-adic numbering system labels all the connections, but not the weightings among the connections.

These weightings are recorded within our mind's living structure, our brain.

Using the mind requires a brain, a preexisting informational structure for the mind to use.

The mind uses the brain to test out possible next thoughts, and, based on the results of this testing, to select the next thought to experience, tending toward next thoughts which maximize information content.

A model for all of this—Digital Mind Math's core quantum process of possibility generation, followed by selection guided by information maximization, followed by experience of the selected thought—is provided by a familiar equation for how quantum physics can be understood to proceed.

We will look at two interpretations of quantum reality—the Copenhagen interpretation and the many worlds interpretation—to see what light they can shed on Digital Mind Math's core cognitive processes.

Schrödinger's Equation

The Austrian physicist Erwin Schrödinger developed a formula early in the twentieth century for the entire set of possible step-by-step ways in which a physical process may proceed, a mapping of the set of all possible step-by-step future evolutions of a physical process.

Schrödinger's equation is intended to model physics at the quantum level, such as the physics of electrons or the physics of photons of light.

The Copenhagen school of physicists—Niels Bohr and associates—used Schrödinger's equation as the basis for its understanding of quantum reality as an unexperienced set of possible paths that, upon observation or measurement, are reduced to a single path that is the path that is experienced.

In the Copenhagen interpretation, the paths that are not selected for experience are not real paths, but are instead hypothetical possible paths that reside in a unexperienced configuration space of what could have been selected.

Here there was an early divergence of interpretation between the Copenhagen interpretation and the many worlds interpretation, for which the unexperienced possibilities reside in a real other world which is a different world than the world of the selected path.

The idea of many other real worlds alongside our own world but inaccessible to our own world may be a bit difficult to accept, but it has proven to be a useful model to help interpret the meaning of quantum reality.

Both the Copenhagen model and the many worlds model add a very special interpretation to Schrödinger's equation: They both use Schrödinger's equation not only to provide all possible paths of a quantum event; they also use Schrödinger's equation to provide the likelihood (probability) of each quantum outcome at each successive quantum step.

In other words, Schrödinger's equation specifies all possible quantum paths, and also provides the probability that, upon observation or measurement, the quantum possibilities will collapse to each particular possible path.

Using this quantum interpretation to understand Digital Mind Math, we find that p-adic mathematics gives us an excellent numbering system to list the entire set of possible next steps. But this use of p-adic mathematics leaves us with two questions: How does our mind get its list of possible next steps? And how does the mind introduce the reality that some possible paths are well-trodden and likely, whereas other possible paths are tenuous and highly unlikely and theoretical?

Here we see why a mind needs a brain. Here we see the importance of the prior history of thoughts that have taken place within our brain.

Each prior thought has shaped and reinforced the pathways within our brain. When our mind uses our brain to generate possibilities for our next thought, it uses the cognitive framework that has been etched into our brain as a result of our entire lifetime history of prior thoughts. The analysis of which next thought to experience takes advantage of this preexisting framework of connections with varying strengths established and shaped by our past cognitive history.

Digital Mind Math's core processes of possibility generation and selection of the next thought resemble the Schrödinger's equation's framework of all possible quantum configurations, each with a probability associated with it. Digital Mind Math leads to the experience of the selected thought by applying its cognitive structure to the individualized brain that has been created from its entire history of past realized thoughts.

Toward Infinite Knowledge

Some conceptualizations of attaining enlightenment through meditation require sustained concentration without actually having a specific thought. It is this uninterrupted harmony with the universe that leads to a path of heightened understanding. It is critical that concrete realities even as amorphous as an experienced thought not interrupt the path toward understanding.

Such conceptualizations offer an interesting application of the core quantum cognitive process, in the following sense.

By avoiding collapse to a single experience—avoiding even the mind's experience of a specific thought—the mind is able to continue wordlessly toward a higher understanding, toward the far-reaching depths of consciousness, toward being one with the universe.

The Digital Mind Math model expects continual collapse to a single experienced thought as the path toward increased information content. The sustained avoidance of this collapse through meditative techniques offers an unexpected application of this model.

What This Chapter Says About Digital Mind Math

There are other ways that quantum physicists interpret reality and interpret Schrödinger's'equation. In fact, Schrödinger himself was not comfortable with others' interpretations of his equation that allow for the possibility that a quantum particle can be both in one location and in another location at the same time (many worlds), or interpretations which postpone until an observation or measurement the realization of a quantum particle's location. Schrödinger hypothesized a rather gory thought experiment—Schrödinger's cat—in which a cat may or may not have perished in a closed box due to poison gas. Schrödinger used this example to illustrate his discomfort with quantum interpretations that deny that the cat has surely already either lived or died—without regard to whether or when the lid was opened and the cat was observed.

Remembering that this is not a book of quantum physics, and remembering that the purpose of Part Three's anecdotes is to gain practice with the various aspects of the mathematically complex Digital Mind Math framework, let's just accept that we have not in this chapter resolved any great questions regarding interpretations of quantum physics. What we have done, however, is show that the mathematics of physics' quantum jump—however this phenomenon is philosophically and physically interpreted—resonates with the dual real and complex $M^4_+ \times CP_2$ mathematics and p-adic mathematics of Digital Mind Math.

The next chapter takes us far away from quantum physics, to completely ordinary places that our minds take us every day.

30. Thinking, Considering, Deciding, Planning, Organizing, Wishing, Regretting, Conversing

THE ORGANIZATION OF THE MIND
- Grouping details to form concepts
  - Linking concepts
  - $M^4_+ \times CP_2$ structure
    - P-adic structure

THE CORE COGNITIVE QUANTUM PROCESS
- **Possibility generation: Generation of possible next thoughts**
  - **Adding detail**
  - **Migrating to a connecting thought**
    - **Selection: Selection of the next thought**
  - **Guided by the maximization of information content**
  - **Negentropy maximization is the only value**
    - **Experience: Experiencing the selected thought**

IN THIS CHAPTER:
Our everyday thought processes can be can be efficiently and effectively modeled as core Digital Mind Math cognitive processes.

Thinking

When we think about something, this means that we complete the cycle of the three-part core cognitive process by experiencing this thought, focused on this "something," this topic. This topic now becomes the last thought that we experienced, and it is from this topic that our mind begins the cyclical core cognitive process again, generating possibilities for the thought that will follow.

This topic—the topic of the last thought that we

experienced—is about "something," and has a subject, a p-adically largest level of understanding, a topic sentence. But the topic is not just its topic sentence; it is an interconnected network of related information within which the topic exists in the context of our entire cognitive organization.

When we experienced this last thought—this topic that, after experiencing it, became the starting point for analyzing what our next thought will be—we experienced the essence of this thought and all of its connecting thoughts.

We now entertain possibilities for what to do with this last thought, how we will construct a next thought that starts with the prior thought and proceeds to a next thought.

Taking advantage of the great efficiency of p-adic mathematics' modeling of our starting topic, we generate many possibilities for what our next topic will be.

Piaget offers a useful framework for the cognitive possibilities: We can either assimilate or accommodate. Our mind generates many assimilating possibilities and many accommodating possibilities.

Next thoughts found by assimilation will keep our focus on the topic that we're starting with, but will add new information to that topic.

To create assimilating possibilities, we insert additional information into the starting topic. We find this additional information in the world external to our mind, or by following links within our mind from our starting topic to wherever these links take us.

Keeping in mind that the p-adic norm is quantified as p^{-n}, the assimilation process increases p. Assimilation means that an additional fact can fit in nondisruptively to the structure of the starting topic, without changing the structure, only with an additional point or additional points of data.

When we find our next thought by Piaget's process of assimilation, we—at one or more places, at the topic-sentence level and/or any p-adically smaller level—increase the p-adic size of these levels by increasing the number of points that this level contains.

Now there are a couple of asides that we need to mention here, in order to keep us mathematically rigorous:

- First of all, remember that p is a prime number—2, 3, 5, 7, 11, 13, 17, 19, 23, and so on. So if our topic at its p-adically largest level had 19 aspects to it (counting 0, the null possibility, as one of these aspects), and if we added two more elements, we have in fact increased the value of p needed to cover this topic from p = 19 to p = 23, even though we've added only 2 elements. But if at the thought following this we add one more element, bringing the total to 22 elements, we still cover this with p = 23, and have not created a need to increase p. In either case, we have increased information content , more obviously when we added elements that increased p from p = 19 to p = 23, but even when we increased the number of elements from 21 to 22, not triggering a higher p. The reason for this is the long-term contribution to information increase that any of this new information provides. The analysis of information maximization is a long-term analysis—a present value of future information content, rather than an examination of only the immediate short-term impact.

- A similar point is to be made about information assimilation that adds elements at informational levels that are p-adically smaller than the topic-sentence level of information. Remember that the p-adic norm p^{-n} depends only on the topic-sentence level's p. So again, there will not be an immediate short-term increase in p^{-n} by adding elements at an informational level below the topic-sentence level, just as there was not an immediate short-term increase in p^{-n} when adding elements that did not bring us to the next higher prime number p. But in either case, we have enriched the organizational structure of what we know, making it more likely in future thoughts to add elements to bring us to a higher prime number p.

Assimilation, like accommodation, does not necessarily immediately, in just one step, at every thought, bring us to a stable higher-level equilibration, to a satisfactory information-maximizing new cognitive organizational structure. For either assimilation or accommodation, there is an expectation of a multi-thought process taking place before a stable equilibration is reached. The processes of assimilation and accommodation are processes of adding information, of cleaning up organizational structure, and of reaching mastery of more complex subjects, subjects with more levels of detail.

Assimilation is a more straightforward—easier to describe— approach to finding our next thought than accommodation is. There are more varied possibilities for how accommodation may proceed, since a next thought that derives from accommodation can restructure our informational organization in several different ways.

As we try out next thoughts by accommodating additional information that was not incorporated within the starting thought, we may, for example combine two or more exisiting concepts. Or we may split a concept into multiple concepts. Or we may switch our focus to a concept that we reached via a CP_2 tunnel, a portal that p-adic mathematics has labeled for us, a portal that relocates where "I" is for the next experienced thought.

Systematicaly, the mind, through its evolutionarily adaptive processes, adds and groups details and links concepts—generating possibilities, selecting, experiencing—increasing information, increasing organization, increasing negentropy. We know when we have more cognitive work to do before we've mastered this topic, before we have a stable understanding. And we may move on, satisfied, when we reach what feels to us like a stable equilibration, a stable resolution.

But as long as we're alive, stable resolutions are only temporary. We urge ourselves to move forward, to continue on. More information. More organizational efficiency and structural integrity. Higher levels of negentropy.

Considering, Deciding, Planning, Organizing

When am I going to plant the garden? How am I going to complete my thesis? Who will we invite over for Christmas?

Many times a day, we create and implement plans, we consider alternatives, decide on a path, and organize for success.

What should I wear today? Which movie should I go to? What will we have for dinner?

Short-term tasks, accomplished successfully many times each day, are completed detail by detail, step by step, process by process. We put on clothes, see a film, eat a steak.

We generate possibilities for these short-term tasks, select, and experience.

They are completed, and become part of our informational structure. Not always an important part, but a part nonetheless.

Maybe nobody commented on what we wore, maybe someone complimented us, maybe someone criticized us.

Maybe the movie or dinner was forgettable, maybe it changed our life for the better, maybe it changed our life for the worse.

Completed events that we've experienced in the past become part of our cognitive informational structure, along with all of their physical and emotional links to other events and along with these other events' physical and emotional links. Sometimes these events are completely assimilated into the existing informational structure of our mind, and sometimes our informational structure accommodates these events by changing the organization of our mind.

Everything we ever do becomes part of our past history of cognitive quantum jumps. The informational structure at each moment reflects our entire past history, and becomes the structure from which our next experience will emerge.

And longer-term events which we consider and plan remain within our informational structure as incomplete thoughts, details without clear boundaries, until we, over the course of days and weeks, decide and plan and execute and experience.

Wishing, Regretting

How could I have said that? Or done that?

Why didn't I speak up? Or keep quiet? Or say yes? Or say no?

If only I could live there, or have that job, or that spouse. . .

We are quite adept at examining past situations, and at hypothesizing alternative or future situations. We do this by starting with an existing concept—a past event that actually happened, a current situation, a practical plan, an idea for the future that we've previously contemplated—and adjusting a detail here, or a relationship there.

We are able to recall or to envision whole M^4_+ events, and to replay them as often as we'd like, with details adjusted and events reshaped in our quest for psychic comfort, for a better life.

The mind—built with p-adic mathematics—easily allows substitutions and restructurings, allowing us to wish for a better life by picturing different scenarios for what was, and what is, and what will be.

And in our melancholy moods of regret, we play back incidents with varying details, and we project forward the sad trajectory that we're on.

How lucky we are to have in place this organizational structure that can modify and regroup details and interconnections, generating and experiencing possibilities in our attempt to be our best.

Conversation

What does it mean for two people to converse, to have a conversation?

Modeled with Digital Mind Math, a conversation between two people is essentially the same phenomenon as thinking, except that there is an alternation between the two people as to which of the two is proceeding with the three-part core cognitive process: In conversation, at each cycle of the core cognitive process, the experience that ends the cycle with one person's comment becomes the starting point for the conversational partner.

In other words, a simple conversation alternates, from one person to the other, which of the two is performing the core cognitive quantum process of possibility generation, then selection, then experience.

Let's examine this more closely.

Possibility generation and selection in conversation. In private thinking, one generates possibilities for the next thought after experiencing one's own prior thought.

In conversation, Person B generates possibilities for the next comment after hearing Person A's prior comment.

For example, suppose Person A just said: "I'm thinking of getting a new car."

Person B now generates possibilities for what to say in response, with the two general categories of possible response being: adding details, or migrating to a connecting thought.

Adding details stays on the topic of Person A getting a new car. Perhaps most obviously there are many possible next comments about what kind of car, the details of the new car: Ford? Toyota?

BMW? Or perhaps: SUV? Convertible? Electric? Hybrid? Or maybe a question about the acquisition process: Which car dealer? Buy or lease? Will you be trading in your old car? How quickly are you planning to proceed with this?

If answered directly, these details will simply increase the value of p for the topic of A's purchase of a new car.

Or perhaps A will have something more elaborate to say about these details, opening up a whole new topic at a more specific level of detail—a higher value of negative-n.

Then there are conversational possibilities that make a connection from Person A's original comment, rather than adding a detail to Person A's comment. Perhaps Person B comments about her own or a mutual friend's recent car-buying decision. Or about the option to get rid of her car in favor of public transportation and Uber.

So many varied comments are possible! And just the briefest moment to select which comment to make!

The selection—from the generated set of possible comments—of which comment Person B will actually make is, as for private thought, guided by information maximization. But in conversation, there are two people's informational structures to consider.

If this is a work situation in which Person A is the boss, then it is probably best for Person B to measure information according to standards relevant to Person A. Perhaps the vehicle is being purchased for the business, or perhaps B is A's personal assistant, responsible for implementing A's decisions.

But maybe A and B are friends, in which case, in service of the friendship, each has the right to get something she wants from the relationship, as well as the responsibility to contribute something to her friend. So in this case, information maximization reflects both A's and B's needs and desires.

Or maybe A is B's son. A is about to get his driver's license, and what A is actually bringing up is Dad lending money to purchase the car. And the insurance. In this case, Person B's needs and desires will weigh heavily in how information maximization is defined.

Or maybe Persons A and B have never met. They're waiting for the subway to arrive, and are both looking in the direction of a billboard for a car dealership. In this case, there is an element of social norms that defines what makes a good conversation. So B will move the conversation only gently in her response: "It's tempting, isn't it, when the subway's delayed." Or "It would just be too expensive for me. And we're so lucky to have the subway."

<u>Construction of a persona</u>. Perhaps we'll use the conversation as an opportunity to construct an image of who we are, who we appear to be or want to be. Our goals and methods are defined and constrained by the situation and the relationship, but conversational interaction gives us an opportunity to select what we say and how we say it, to build an image, to eliminate or de-emphasize a detail here, or shade a detail or tell a little white lie there.

In conversation, there are opportunities and risks presented by the fact that each cycle ends with the experienced thought or comment becoming someone else's, not one's own, starting point for the next cognitive cycle.

Conversation can be thought of as two people thinking out loud: The processes of thought take place in one person's mind, but the art of conversation involves two (or more) people attempting to have a shared experience of the core cognitive quantum process.

What This Chapter Says About Digital Mind Math

The core cognitive process can be applied straightforwardly to a topic, monotonically increasing information content along a well-defined rigid path.

Or we can toy with the cognitive process, have fun with it, be tortured by it.

Ultimately, we are driven by the urge to maximize negentropy and minimize disorder. But the incremental steps and tactics in service of this ultimate strategy are quite varied and dependent on individual history.

This is what makes life the interesting and challenging journey that it is.

In the next chapter, we continue examining how it is that we tend toward the best that we can be.

31. Creativity and Innovation

THE MECHANICS OF DIGITAL MIND MATH

THE ORGANIZATION OF THE MIND
- Grouping details to form concepts
  - Linking concepts
  - $M^4_+ \times CP_2$ structure
  - P-adic structure

THE CORE COGNITIVE QUANTUM PROCESS

- **Possibility generation: Generation of possible next thoughts**
  - **Adding detail**
  - **Migrating to a connecting thought**

IN THIS CHAPTER:
Innovation—thinking outside of the box—is the process of finding new creative solutions. Digital Mind Math has a built-in core process for facillitating this: the process of generating possibilities.

In this chapter, we focus in particular on how especially creative techniques for adding details and making new connections can lead to disruptive innovation.

- Selection
- Experience

Everyday Creativity and Play

We've discussed learning and understanding and making sense of the world as incremental processes, in which information increases steadily, in increments of moderate novelty. Consistent with this framework—Piaget's framework of moderate novelty—we have Vygotsky's tools of the mind and intentional play, targeting a zone of proximal development for optimal cognitive as well as social and emotional development.

During play, the child is absorbed in the moment. The Argentine writer César Aira calls this the "perpetual present": "The immediate absorption of reality, which mystics and poets strive for in vain, is what children do every day."[76]

This pattern continues into adulthood, as we continue to play games as a sort of p-adic comfort food, and as we watch movies or we read fiction or we gossip for cognitive or emotional stretching. And in any form of studying or working or learning—even though it no longer feels like play—we can still feel energized and derive a motivating sense of satisfaction from information-increasing jobs well done. We at times lose ourselves in our games and other leisure activities and in our work.

And play isn't just for fun. Play is "a disciplined make-believe that leads to *ekstasis*, a 'stepping outside' of normal perception, which, when translated into action, has also helped to develop law, commerce, art and science." And we have a "human propensity to play."[77]

We are driven to play because we are driven to explore, to generate possibilities, to increase information. We have a "hungry mind," a natural sense of curiosity.[78]

So we have a natural baseline of creativity built into our everyday cognitive processes. We are naturally curious. We hypothesize and we play. Throughout the day, every day, these experiences increase information and order.

But what does it take for true outside-the-box thinking and innovation, for novelty that is not moderate but disruptive?

Disruptive Innovation

How does innovation differ from everyday creativity?

Quantitatively, it can be said that innovation is more extreme than everyday creativity, more intensive, accomplishing more. It is more adventuresome, more exploratory.[79]

But there are qualitative differences also. Innovation is divergent, rather than convergent. It is multidisciplinary, drawing from varied fields. It often emerges as eureka moments at bath time or nap time, while driving or exercising—as random episodic silent thought (REST).[80]

How can we make sense of this within the Digital Mind Math model?

The characteristics of disruptive innovation, of genius, seem to relate most to the process of possibility generation, the process of generating possible next thoughts.

The highly innovative experience derives from unusual ways of adding details and of making new connections. Disruptive innovation—the work of genius—is associated with drawing details from multiple sources and disciplines, and with making connections that are extraordinary, not typical.

An intriguing aspect of neurological study is the role of chaos within the nervous system and the brain. First a definitional clarification: Chaos is not the same as disorder, and in fact is something of an antonym to disorder. Chaos relates specifically to large effects caused by small changes in input. So chaos is not disordered. A chaotic outcome is deterministic and ordered. But small tweaks can cause wildly varying effects.

There is some current theorizing that chaos is an important natural aspect of neurological functioning, which helps create a "a surprising harmony, in so-called chimera states."[81] Perhaps it is the wide sourcing of details and the unusually creative connections that lead to disruptively innovative outcomes, the creatively imagined hybrid animal of myth and magic, the chimera.

And let's not forget the role of perspiration. Thomas Edison is often cited as crediting genius 99% to perspiration and just 1% to inspiration. So innovation is correlated with a bias for action, with experimenting with many possibilities, with experiencing a wide selection of thoughts. Perhaps this is part of what Shonda Rhimes,the wildly successful writer and producer of popular televison shows, means by *Year of Yes*: By making a point of saying "yes to all the things that scared me, that made me nervous, that freaked me out, that made me think I'm going to look foolish doing it," Rhimes took herself out of an introverted comfort zone into a world of bold creativity.[82]

What This Chapter Says About Digital Mind Math

Exceptional creativity and disruptive innovation may result from ordinary cognitive processes that are exceptionally executed. In particular, this chapter focuses on the twin aspects—adding detail and creating connections—of Digital Mind Math's core cognitive process of generating possible next thoughts. Disruptively innovative possibilities draw from wide sources and find rare connections.

32. Shared P-Adic Structures: Religions, Philosophies, Memes, Languages

THE MECHANICS OF DIGITAL MIND MATH

THE ORGANIZATION OF THE MIND

> • **Grouping details to form concepts**
> • **Linking concepts**
> • **$M^4_+ \times CP_2$ structure: Spacetime 4-spheres, intricately linked**
> • **P-adic structure: Sequences of numbered 4-spheres and their connections**

THE CORE COGNITIVE QUANTUM PROCESS
• Possibility generation
• Selection

> ○ **Guided by the maximization of information content**
>
> ### IN THIS CHAPTER:
> Cultural memes and beliefs are informational structures—at a shared, societal level—subject to the same organizing principles that are at work in organizing information within the mind. Each individual adapts these shared p-adic structures to create private internalized working versions.
>
> Widely available instantaeous online networking pulls us in two different directions, allowing a hyper-personalization of choices that move the modern life away from some shared societal values, at the same time that new social connecting forces are established.

• Experience

Humans are a sociable lot and share a number of ideas and values—for example, religions and philosophies. These culturally shared ideas and values require individuals to learn about them and to internalize knowledge about them. Religions, for example, generally have a core set of guidelines and beliefs, and individuals learn about the tenets of their religion and internalize their religion as a way of life.

Like any kind of knowledge, core religious beliefs may me understood as a Digital Mind Math set of information, dually mapped as a p-adic structure and a real (technically: complex) $M^4_+ \times CP_2$ structure. Details are grouped to form M^4_+ concepts, and concepts are intricately linked in CP_2 space, all efficiently mapped p-adically as sequences of numbered 4-spheres and their connections.

The various aspects of the system of beliefs that define a religion may be examined in the same way that we have, throughout Part Three of *Digital Mind Math*, examined the various aspects of the organization of the mind. The same is true for other socially shared information, such as memes that spread within a culture, and such as the languages that are spoken and heard and written and read.

Each of these phenomena—religions, philosophies, memes, languages—represents a set of information which can be understood using the same organizational principles as we've been using to understand the organization of the mind. And these organizational principles include not only the organizational model of the mind, but also the core Digital Mind Math organizing principle: that they emerge as a natural human process of life, tending toward information maximization and in opposition to disorder.

A striking aspect of these culturally developed phenomena is that they become internalized within individuals. Languages, for example, have a shared p-adic structure that individuals must internalize. Each speaker of a particular language creates within her mind the intricate structure of her native language.

So language has a shared p-adic structure, and also a p-adic structure that each speaker internalizes. There are many variations even of the shared p-adic structure—a complete unabridged dictionary of words and their meanings, a polished set of usage rules, and many regional and cultural gradations. And certainly there are variations in each individual's structure of their language—variations in style and in completeness and even in correctness.

Every language has a p-adic structure that is socially shared, and individuals create their own versions in their own minds as the p-adic structure which they access to read, write, speak, and understand.

Social Media: A Culture of Narcissism? Or Human Swarms and Flocks?

Even before the explosion of the new social media culture, social historian Christopher Lasch struck a chord when he published *The Culture of Narcissism* in 1979, citing the decline of family and other social traditions as contributing to heightened individual self-focus and egocentrism. [83] Since then, the creation and maintenance of one's digital self has taken over much of the time and energy of many. Business schools and marketers are fully cooperating by helping businesses to harness the power of data and analytics in order to hyper-personalize their products and their customers' experience.

But is this only narcissism, or is this our path to new visions of human cooperation? Is this loss of autonomy not a threat to be feared, but rather the beginning of a new era of worldwide socially shared consciousness?

"Humans are by far the most social species, other than insects."[84] So we're just second best? Swarms of bees or ants, or perhaps even flocks of birds in flight, have greater shared social p-adic structures than we do?

Maybe the whole social media constellation—Facebook, Twitter, Instagram, Tumblr, LinkedIn, and all of the new and emerging media to follow—are driven by an emerging social cognition and our urge to heighten as yet untapped capabilities for maximization of shared human information.

What This Chapter Says about Digital Mind Math

To function as a society, we have many social institutions, each with their own rules, patterns, and informational content. Written, audio, and video materials document the content and organization of this information. Experts study, debate, and continually refine the information. Popular advocates have media platforms.

These social institutions are kept alive by individual adherents internalizing the specifics and accessing these specifics regularly as a way of life. Typically, adherents will have expertise in a version with streamlined content and interconnections.

The Digital Mind Math structure and process apply at both the social and individual levels.

33. How to Solve Personal and Interpersonal Problems

THE MECHANICS OF DIGITAL MIND MATH

THE ORGANIZATION OF THE MIND
- Grouping details to form concepts
  - Linking concepts
  - $M^4_+ \times CP_2$ structure
  - P-adic structure

THE CORE COGNITIVE QUANTUM PROCESS
- Possibility generation: Generation of possible next thoughts
  - Selection: Selection of the next thought
  - Guided by the maximization of information content

  - **Negentropy maximization is the only value**

  IN THIS CHAPTER:
  Maximization of information content is the driving force within Digital Mind Math and within cognition. So if the question is how to solve personal and interpersonal problems, Digital Mind Math's answer must be: by increasing information content.

- Experience: Experiencing the selected thought

How do we make ourselves a better person?

If we're looking to Digital Mind Math for an answer, this answer must be: Every chance you get, and as often as you can, decrease your cognitive entropy, increase order, increase information content as measured by the p-adic norm.

You know how to do this: Increase p. Or increase negative-n.

To be a better person, add bits of information to your life—by reading, by listening to others. And organize this information, make connections, synthesize.

Do this as often as you can, at every cognitive moment.

After all, isn't this how mediation proceeds, whether it is between warring spouses or warring nations? First, you separate the warring parties and get each to identify what they want, what they don't want, what they hate about their opponent, maybe even what they like, or have ever once liked, about their opponent.

The mediator looks for any common threads, and looks for any possibilities for small reconciliations. Then the hard work begins: a statement of the irreconcilable differences, and—one by one—a statement of the two parties' positions.

Michael Graziano, professor of psychology and neuroscience at Princeton University, and author of *Consciousness and the Social Brain*, draws on his own research, and the work of philosophers such as Daniel Dennett and Patricia Churchland, to develop a radically modern concept of what consciousnes is and what the mind is. He views this question as the third of three fundamental questions of modern science, the first two of which have already been answered: We are not at the center of the universe (established by Copernicus in the sixteenth century). And we are not a qualitatively unique form of biology (Darwin, ninetenth century).[85]

Graziano's third question is: Are we really conscious? And his answer is: No, not as we usually understand this question. "There is only information in a data-processing device," he says. What we possess is attention—the mechanistic ability to focus, to enhance some signals. But awareness—consciousness—is a "cartoonish reconstruction of attention."[86]

There is only information. This is a modern and radical conceptualization of the mind. And also a modern and radical conceptualization of physics.[87] And also the core of Digital Mind Math.

What This Chapter Says About Digital Mind Math

Each cognitive moment presents an opportunity to increase information content, to become more ordered, to create a higher level of equilibration.

Our biology drives this process. Life drives this process. This is the definition of life.

Life is the drive that opposes the relentless increase in entropy that the Second Law of Thermodynamics tells us is the law of the physical world.

This drive—to maximize negentropy—is our life force.

PART FOUR: THE IDEAL MIND

One might argue that it is completely outrageous to claim that a mathematical model of the mind can be developed. Yet this has been the exact argument of *Digital Mind Math* so far:

- The mind operates according to p-adic mathematics.
- P-adic mathematics is the natural mathematics of cognition, having developed through natural evolutionary processes due to the inherent advantages of p-adic mathematics:
    - P-adic mathematics gives us a simple way to capture, analyze, and manipulate enormous quantities of information.
    - P-adic size is a natural measure of information content.
    - The p-adic approximation methodology, as defined through Hensel's Lemma, is *"guaranteed to converge."*[88]

This guarantee to converge is important because it allows us to make decisions that increase informational content on a local, short-term, immediate basis, and upon doing so, to be guaranteed that we have increased informational content on a global, long-term basis.

This may not sound like a great advantage, but for one thing it's not a feature that real mathematics (or its all-encompassing cousin, complex mathematics) offers. If our mind operated using real or complex mathematics, and if we made a good, effective, information-increasing local decision (as measured by some real or complex metric), there would be no guarantee that this decision would be a good one for the long term.

The mathematical approximation process frequently used in real mathematics to find answers to mathematical questions is called Newton's method. This method proceeds by making a guess at the answer, then, according to a formula originally developed by Isaac Newton, to use that guess to systematically get closer and closer to the answer—to converge to the root.

When it works, Newton's method, applied to real mathematical polynomials, converges surprisingly rapidly, and it is a popular computer science algorithm for solving mathematical problems.

The disadvantage of Newton's method is that it does not always work for real mathematical problems, because—bizarrely and problematically—it requires the initial guess to be sufficiently close to the final answer. This is quite a disadvantage, of course, since we don't know the final answer.

The fact that this requirement of initial closeness does not apply to the p-adic Newton's method offers p-adic mathematics an evolutionary advantage in becoming the natural mathematics of cognition. This claim of evolutionary advantage for p-adic mathematics is a contention of Digital Mind Math, a contention which sits alongside the claim that p-adic mathematics also offers advantages in organizing, analyzing, and measuring information.

But let's look a little more closely at exactly how mathematicians describe the advantage of the p-adic Newton's method over the real (or complex) Newton's method. It's still enough to give p-adic mathematics its evolutionary advantage, but there is nevertheless an important qualification. Read closely:

> "In one respect . . . Hensel's Lemma is better than
> Newton's method in the real case: in the *p*-adic case
> the convergence to a root of the polynomial is
> guaranteed by universal conditions on the

approximate solution . . . whose form does not depend on the polynomial. In the real case, Newton's method converges if the approximate solution is sufficiently close to the actual root, but the condition of closeness depends on the polynomial."[89]

With the exception of one aspect, the entire paragraph above is simply a mathematically precise statement of the more casual discussion that we engaged in before quoting the paragraph. The exceptional aspect—the important qualification that we alluded to— has to do with the "universal conditions" that guarantee p-adic convergence.

Now let's be clear: Universal conditions are better than conditions that "depend on the polynomial." In fact, it is the contention of this Digital Mind Math argument in favor of p-adic mathematics that conditions that depend on the polynomial (which is the best that real mathematics can offer) are disqualifying, whereas we can work with universal conditions that do not depend on the polynomial.

In other words, real mathematics evolutionarily failed as the mathematics of cognition because decisions that are good decisions measured short-term using real mathematics cannot be guaranteed to be long-term good decisions, unless we know that these short-term decisions are good long-term. This wouldn't work: It's circular logic, and it doesn't get us anywhere. And it's not a problem when we use p-adic mathematics as the mathematics of cognition.

But even p-adic mathematics will work—will permit good local, short-term decisions to guarantee good long-term results—only if the short-term decisions meet certain universal conditions. By universal conditions, we mean conditions that don't require us to know the answer that we're trying to converge on. Rather, they are conditions that we can identify based on the question.

In Part Four, we will take great advantage of these universal conditions: We will use the universal conditions to identify:

- The ideal way to think
- The ideal way to interact with people
- The ideal way to engage in conversation
- The ideal way to act
- The ideal way to live your life

Yes, this is a completely outrageous claim: There is a mathematical formula that tells us how to think, live, converse, interact, act.

We find this mathematical formula in the universal conditions that tell us when Hensel's Lemma works.

An Example

Suppose we've been invited to a party. We're not wildly enthusiastic about going, and in fact are inclined to skip it. Other possible ways to spend the evening seem more promising. However, we don't want to hurt the feelingsof the friend that invited us. So our thoughts, and our text messages, proceed as follows:

What we're really thinking: Who's going to be there? I don't feel like really dressing up. Will it be casual? It's been a while since I had some good food. Is this going to be crackers and dip, or something tastier? How am I going to get there and home?

What we text: That party Thursday sounds great. I didn't know they had room for all those people. Are you driving?

Friend: They have parties all the time. Real fun group. I'll pick you up at 8:00?

What we're thinking now: Hmm. Sounds like it might be promising. Maybe I'll go, but still not 100% convinced.

What we text: Great. Will you have dinner before you go? Just wear what you wore to work?

Friend: Don't eat anything before. Food will be great. Work clothes or more casual.

What we're thinking: Well this worked out. I'll just record my TV shows. Leftovers stay in the freezer. And it never appeared that I had any second thoughts about accompanying my friend.

What we text: OK

We did not know the answer as we started out. We needed to figure out a way to dance around the issues that were on our mind, without jeopardizing our friendship, and without being explicit about our baser selfish motives regarding comfort, convenience, and the quality of food and people.

So, as we've learned to do, we converge on the optimal solution by a tried and true universal methodology that does not depend on the actual issues at hand.

We increase information content subject to a protocol that converges rapidly, and that—based on our accomplished social skills—we know guarantees finding the ideal long-term solution as long as we increase information step by step, locally, short-term.

Converses and Inverses

For most of this book, we've worked with the formulation that, according to Hensel's Lemma:

IF we use p-adic numbers,
THEN good local decisions guarantee good global decisions.

We now know, based on the introductory comments to Part Four, that we need to refine the IF condition a bit. The more accurate statement of Hensel's Lemma is:

IF (1) we use p-adic numbers, and (2) we satisfy certain universal conditions,
THEN good local decisions guarantee good global decisions.

Now we want to take this even further: We want to tighten the basic version of Hensel's Lemma so that we so specifically and narrowly define the universal conditions (referred to in part (2) of the IF clause) that, if we don't meet these tightened conditions, then we don't guarantee good global decisions.

This is the gold standard for identifying the evolutionarily efficient mathematics of cognition. This is what tells us exactly how to focus our short-term decision-making, so that it guarantees good global decisions *with no waste*—so that it guarantees good global decisions in the most narrowly specified, efficient way possible.

First, let's remind ourselves of the motivation for doing this: This is what will give us the mathematical formula for how best to think, interact, converse, act, live. We can get away with operating a bit more sloppily or broadly than these narrow conditions specify, and still guarantee that local good decisions will produce global good decisions. But, if we want to be all that we can be, we don't want to waste effort, we don't want to be unfocused, we don't want inefficient habits—we want to hit the sweet spot of the most narrow formulation of next steps that will get the job done, that will guarantee long-term success.

Said another way, we are looking to establish the converse, or equivalently the inverse, of Hensel's Lemma: We want to know (the converse) that, if we are guaranteeing good global decisions, we've done it in such a way that we met specific tightened conditions. We want to know what is the bare minimum that will always be true—that must be true—if we are guaranteeing good global decisions.

Or equivalenly (the inverse), we want to know that if we haven't met these exact tightened conditions, we are not guaranteeing a good global decision. We want to know the exact target to meet for condition tightening in order to guarantee the consequence of a good global decision. We are fully aware that we can achieve good global decisions by looser conditions, but this is not what we want. We want to be efficient; we want to live a focused life.

Let's be clear and say this another way: If all we do is abide by broad universal conditions, it is true that we guarantee good global decisions, but not because we needed all those broad conditions—rather, because the broad conditions included the more narrowly focused specific conditons that were the bare minimum required.

Mathematically or logically stated, if we know a basic conditional truth IF X, THEN Y, this is not necessarily equivalent to the converse IF Y, THEN X.

For example, suppose it's true that IF today is Tuesday, THEN I have to go to school. Does this automatically make the converse true: IF I have to go to school, THEN today is Tuesday?

Probably not.

Not if we go to school five days a week, Monday through Friday.

So a conditional statement being true does not necessarily mean that its converse is true. IF X, THEN Y does not necessarily mean IF Y, THEN X.

Nor is the inverse necessarily true. It is not necessarily true that: IF today is not Tuesday, THEN I don't have to go to school.

Today could be Monday, and I would still have to go to school.

IF X, THEN Y does not necessarily mean IF not X, THEN not Y.

Mathematically and logically, certain variations on a basic conditional statement are always true, and certain variations are not necessarily true.

Said another way, if a basic conditional statement is true, then its contrapositve (IF I don't have to go to school, then today isn't Tuesday) is true. But the converse and inverse are not necessarily true. However, the converse and inverse are always in synch with each other, just not necessarily in synch with the basic original conditional statement (or its contrapositive):

Always true together: An original conditional statement (IF X, THEN Y) and its contrapositive (IF not Y, THEN not X)
Synonym for IF X, THEN Y: X is a sufficient condition for Y
Another synonym for IF X, THEN Y: X only if Y
Another synonym for IF X, THEN Y: X implies Y; or in symbols: $X \Rightarrow Y$

Always true together: The converse (IF Y, THEN X) and the inverse (IF not X, THEN not Y)
Synonym for IF Y, THEN X: X is a necessary condition for Y
Another synonym for IF Y, THEN X: Y implies X; or in symbols: $Y \Rightarrow X$, or: $X \Leftarrow Y$

Not necessarily true together: The original conditional statement and its converse

Not necessarily true together: The original conditional statement and its inverse

Similarly: If we know that the contrapositive is true, we don't know one way or another whether the converse and inverse are true.

What's motivating this discusion is that we have Hensel's Lemma stated as a basic conditional statement:

IF (1) we use p-adic numbers, and (2) we satisfy certain universal conditions,
THEN good local decisions guarantee good global decisions.

and we want to know if its converse (or inverse) is also true. We want to know that we're making good global decisions in a way that good global decisions must be made—not just in a way that it's possible to make good global decisions. We're looking for the necessary conditions for making good global decisions, not just the sufficient conditions.

We want to know how to live our lives so that, at every moment, we make decisions that most effectively maximize our lifetime informational content.

Tightening Hensel's Lemma

In some circles, Hensel's Lemma is very popular, and in fact there are a number of variations on how it is stated, including variations on the statement of the universal conditions that must be met in order for Hensel's Lemma to hold true.

The bad news is that even the simplest statement of these universal conditions requires a significant level of mathematical understanding. And the more elaborate statements are highly advanced and esoteric.

So here in *Digital Mind Math*, we are able to give only an intuitive formulation of these universal conditions, and we will have to rely on an interested advanced mathematician who also possesses a rare common touch to provide a more precise, yet still accessible, formulation.

What's usually considered the simplest formulation of the universal conditions that guarantee that Hensel's Lemma applies is that the derivative of the polynomial being examined not be zero.

OK. I've lost 99% of the readers.

But this is the simplest version, so let's try to gain some intuition for it, starting with an intuitive statement of what a derivative is: A derivative is a measurement of change. So if the derviative is zero, there is no change. If the dervative is nonzero, there is change.

We're supposing in Digital Mind Math that our minds have learned to test all possible next thoughts for the purpose of selecting a particular next thought that tends to maximize information content. So with our goal being to maximize long-term information, if our analysis is performed p-adically, and if we increase the local p-adic measurement of information content, then we will increase the long-term lifetime information content.

The simplest form of Hensel's Lemma is essentially nonrestrictive. It says that, to know we've continued on the path of lifetime information increase, all we have to do is make decisions that at each moment increase the informational content compared to the previous moment.

We know how to do this, from the earlier parts of *Digital Mind Math*: We can either increase p, or we can increase negative-n. We can either increase the number of details in the topic at hand, or we can put the topic at hand into a more detailed context.

So the simplest formulation of Hensel's Lemma tells us what we've known from the very early chapters in *Digital Mind Math*: If we use p-adic mathematics and if we increase information (measured p-adically) when we select which thought to experience as our next thought, then we can rest assured that we have increased our lifetime informational content. (If you're thinking that this is no big deal, just remember that this statement is not true for real numbers.)

The only problem with this simplest implementation of Hensel's Lemma is that it's too broad. We're wasting effort. It's not the narrowest formulation of what it takes to increase lifetime information content. It's a true conditional statement, but one for which there is no claim that the converse of this conditional statement is true. There is no claim that we've identified a necessary condition, only a claim that we've identified a sufficient condition.

This is not the ideal way in which we want to live our lives. We only have so long to live, and we want to make the most of it. We don't want to settle for what's a sufficient next step, one that we can get by with, one that will work, but may be wasteful. We want to know what a necessary next step is. We want to operate in a region in which the steps we've taken are focused like a laser beam on increasing our lifetime informational content.

"Tightening the Basic Version of Hensel's Lemma" is a brief paper by mathematician Keith Conrad in which he presents a version of Hensel's Lemma that "provides a converse of sorts."[90] The good news is that he has narrowed to the bare minimum the statement of conditions under which convergence to the solution is guaranteed. The conditions in Conrad's tightened version of Hensel's Lemma are conditions that, if we narrowed them any further, we would no longer guarantee convergence. And, if these tightened conditions were any wider, they would be needlessly broad, would be overkill, more than we need, wasted effort.

This is what the converse of Hensel's Lemma gives us: the exact statement of how to live each moment so that we have most efficiently continued on a path toward maximum lifetime informational content, toward the best that we can be.

The bad news is that there are only a few hundred mathematicians in the world who can understand this, and I have not yet met one (in spite of my efforts) that I have convinced to spend the time and effort to translate this into a workable Digital Mind Math framework.

Here's what can be said, as an intuitive statement of the converse of Hensel's Lemma:

- The condition that must apply still involves the derivative of the polynomial being examined—the change in the measured quantity, which for Digital Mind Math is the informational content.
- But unlike the simplest version of Hensel's Lemma (for which the converse is not true), if we are in fact looking for the converse of Hensel's Lemma, then it is not the case simply that the derivative must be nonzero.
- Instead, the magnitude of the derivative must be greater than the square root of the magnitude of the original quantification. This is what Conrad shows is the universal condition for the converse of Hensel's Lemma. Let's now work to try to understand this intuitively.
- Conrad's work is developed in the p-adic region in which we are dealing only with whole p-adic numbers (no p-adic decimals).
  - P-adic mathematicians prefer to say this as: Both the p-adic magnitude of information content, and the p-adic magnitude of the change in information content, are less than or equal to 1. We know that a

p-adic number having magnitude less than or equal to 1 is equivalent to a p-adic number being a whole p-adic number (not a p-adic decimal) because the norm of a p-adic number is p^{-n}. Since n is the location of the last digit to the right, if there are no digits to the right of the decimal point, n can't be a negative number. So the largest magnitude that a p-adic whole number can have is p^0, which is 1, if the p-adic whole number has its units digit filled (for example, 123456). Or the magnitude of the p-adic number is less than 1 if the last nonzero digit is further to the left (for example, 123000).

o This is different from the p-adic region that we usually assumed in developing our Digital Mind Math intuition in Parts One through Three, where we typically operated in the decimal p-adic region, where adding levels of detail means adding more decimal digits to the right. In Conrad's derivation of the universal conditions that tighten Hensel's Lemma to its converse, both the p-adic value, and the change in p-adic value, are whole p-adic numbers. So we'll need to reorient our intuition to this p-adic region so that we can understand what Conrad's work intuitively means

- For our "should I go to the party" example, this means we could model our going to the party on Thursday with our friend as, for example, M^4_+ event number 6034200000 (a 7-adic number ending in 5 zeroes).

- This event number 6034200000 is a big event for us in a real sense. OK, it's not as big an event as say the birth of a child or getting married. But it does involve getting dressed, going out of the house, bonding with our friend, and maybe meeting our future spouse.

- The food at the party is a detail—maybe 6034251000. And whether we're going to change our clothes after work is even more detailed—maybe 6034236240.
- So with the p-adic magnitude being p^{-n} , the p-adic magnitude of the party is $7^{-5} = 1/7^5 = 1/17407$. The p-adic magnitude of the food at the party is $7^{-2} = 1/7^2 = 1/49$. The p-adic magnitude of what clothes to wear is $7^{-1} = 1/7 = 1/7$.
- We know that the universal condition for the converse of Hensel's Lemma—which is stated for p-adic whole numbers—is that the p-adic magnitude of the derivative of the core function is greater than the square root of the p-adic magnitude of the core function itself.
  - o For our example, the p-adic magnitude of the core function is the informational content associated with attending the party. And the p-adic magnitude of the derivative of the core function is the incremental informational contribution of various aspects of the party experience.
  - o Sometimes it's easier to think of the universal condition as: The magnitude of the derivative is greater than the square root of the magnitude of the basic core function. At other times, it's easier to think of the universal condition in an equivalent form: The square of the magnitude of the derivative is greater than the magnitude of the basic core function.
- We just need to look at the exponents in order to see if we're satisfying the condition that the square of the magnitude of the derivative is greater than the magnitude of the original quantification: For the food, $(7^{-2})^2 = 7^{-4}$, which

is greater than (a bigger number than) 7^{-5}. And certainly for changing our clothes after work $(7^{-1})^2 = 7^{-2}$ is greater than 7^{-5}.

- The more general statement is that, if the magnitude of the original quantification has an exponent of $-(2M+1)$, then the magnitude of the derivative must be -M or a smaller negative number.

- This is why when $-(2M+1) = -5$, we're satisfying the converse of Hensel's Lemma when -M = -2, and we're also OK with an exponent of -1.

- Stated in an intuitive way, this is why our highly developed social skills led us to dance around our actual concerns about whether or not to go to the party—our social skills led us to avoid point-blank questions about the food and the crowd, etc.—and instead to draw out the information that interested us by asking indirect questions, with hazier focuses than the party as a whole and even than our full set of actual concerns.

- An original condition of xxxxxx00000 [5 zeroes] must have a derivative no larger in a real sense than xxxxxxxx00 [2 zeroes] in order to satisfy the converse of Hensel's Lemma.

- An original condition with 7 zeroes must have a derivative with no more than 3 zeroes in order to satisfy the converse of Hensel's Lemma.

- An original condition with 11 zeroes must have a derivative with no more than 5 zeroes in order to satisfy the converse of Hensel's Lemma.

- An original condition with 2M+1 zeroes must have a derivative with no more than M zeroes in order to satisfy the converse of Hensel's Lemma.

So this is it—the answer to the central question of religion, of philosophy, of therapeutic psychology: How do we best live our lives?

The answer is that, measured p-adically, at each moment of our lives, we must choose the route for which the square of the p-adic magnitude of change in informational content is greater than the p-adic magnitude of the prior moment's informational content. Or, equivalently, the magnitude of change in informational content is greater than the square root of the prior informational content.

Go forth and lead your life accordingly!

The only thing that needs to be clarified is how to:

- Map out the precise p-adic formulation of a moment's thought, using whole p-adic numbers
- Take the derivative
- At each moment of your life, choose the path that increases information content by more than the square root of the moment's information

We will close Part Four with a number of examples that use the Digital Mind Math framework to show how the converse of Hensel's Lemma is applied in everyday life—even without our being aware that this is what we're doing. In many cases, what we find is that living life according to the converse of Hensel's Lemma is a natural process, the cognitive process that we naturally use as we proceed with life.

In other words, not only (parts One through Three of *Digital Mind Math*) do we naturally think p-adically, we naturally (Part Four) think p-adically consistent with the converse of Hensel's Lemma. This is what is illustrated in the examples that follow.

As a reminder, stated with mathematical elegance, in terms of the derivative, we satisfy the converse of Hensel's Lemma—and therefore live our life to the fullest—if:

$$| \ f(a_0) \ |_p < | \ f'(a_0) \ |_p^2$$

Because it may be simpler to think about the derivative—how we can change events—we'll also want to keep in mind the equivalent statement, but placing the derivative on the left side:

$$| \ f'(a_0) \ |_p > \sqrt{| \ f(a_0) \ |_p}$$

Probably most intuitively, this means that for a specific value of p (that is, in a p-typical environment, rather than an environment of universal p), if our current state is p-adically labeled 120000000 [7 zeroes], then we're looking for a derivative of (change in) our current state that extends to a more refined level of detail in a real sense (which menas that it is larger in a p-adic sense) than 123456000: 3 zeroes, or fewer than 3 zeroes (123456700 or 123456780 or 123456789). This means that the derivative is more detailed (p-adically, with a higher informational content) than the original condition, and is more detailed to a specified extent:

- If the exponent within the magnitude of the original condition is -7, then the exponent within the magnitude of the derivative must be -3 or -2 or -1 or 0 (that is, ending with 3 or 2 or 1 or 0 zeroes).
- If the exponent within the magnitude of the original condition is -6 or -5, then the exponent within the magnitude of the derivative must be -2 or -1 or 0 (that is, ending with 2 or 1 or 0 zeroes).
- If the exponent within the magnitude of the original condition is -4 or -3, then the exponent within the magnitude of the derivative must be -1 or 0 (that is, ending with 1 or 0 zeroes).

- If the exponent within the magnitude of the original condition is -2 or -1, then the exponent within the magnitude of the derivative must be 0 (that is, ending with 0 zeroes).

- More generally, if the exponent within the magnitude of the original condition is -(2M + 1) (that is, if the original condition is represented as a p-adic whole number ending with 2M+1 zeroes), then the exponent within the magnitude of the derivative must be -M or a smaller negative whole number (that is, the derivative must be a p-adic whole number ending with M or fewer zeroes).

It is this last formulation of the converse of Hensel's Lemma—involving the number of zeroes that the p-adic whole number ends with—that offers us the most promising intuitive route for discussing this in the examples below.

But remember that Digital Mind Math is not limited to p-typical p-adic mathematics: In Digital Mind Math, we don't require every level of our thought hierarchy to have the same number p of elements. We operate in a world of universal (not just p-typical) Witt vectors.

So there will be times in the examples below when we'll be referring to the

$$| f'(a_0) |_p > \sqrt{| f(a_0) |_p}$$

formulation of the converse of Hensel's Lemma, since the simplification involving exponents -(2M + 1) and -M applies only in a p-typical environment. Operating in a p-typical environment permits increase in magnitude only by increasing the levels of detail, but we know in Digital Mind Math that the magnitude of information content is also affected (in a universal-p environment) by the size of p, the number of details within a given level.

For our highly intuitive example regarding cleverly figuring out whether to accept the invitation to Thursday's party, this explains why we're best off not bombarding our friend with all our questions at once, but instead asking these questions just a few at a time:

- Our goal is that the p-adic magnitude of the derivative—our questions that inch us toward a decision—is greater than a certain minimum size (that is, greater than the square root of the magnitude of partying).
- Without drowning too much in the mathematical details, we do need to wrap our heads around the thought that our statement of the converse of Hensel's Lemma assumes that we're operating strictly in a region of whole p-adic numbers only (both $| f(a_0) |_p$ and $| f'(a_0) |_p$ less than or equal to 1).
- So, translated from a p-typical statement to a universal-p statement, the condition satisfying the converse of Hensel's Lemma—in the region for which the terms and specifications of the converse of Hensel's Lemma are derived—is that we need to *lower* the derivative's p sufficiently, compared to the original condition's p, in order to ensure that $| f'(a_0) |_p > \sqrt{| f(a_0) |_p}$.
- This is because for p-adic whole numbers, the exponent -n in the norm p^{-n} is always a negative number (or zero). Therefore, in the region in which the converse of Hensel's Lemma is stated, the p-adic norm increases as p decreases.
- This is why optimal social skills involve inching our way toward what we're after by asking just a few, not all, of our questions at once: We want the derivative to be large enough to satisfy the converse of Hensel's Lemma, implying that we need to lower p.

• So in our examples below, we'll be looking for derivatives that have smaller negative-n (fewer zeroes than the original condition) or smaller p (fewer details). These are the ways in which we meet the conditions of the converse of Hensel's Lemma, and therefore live the ideal life.

In Digital Mind Math, we remain committed to the concept that p-adic mathematics is our first mathematics and the natural mathematics of cognition. The mind developed p-adically because of the evolutionary advantages of p-adic mathematics. The following examples will help us understand the aspect of this evolutionary advantage that derives from the converse of Hensel's Lemma, and how this plays out in our everyday life.

Technical Note on Scale, Normalization, and P-Adic Norm Less Than or Equal to 1

As we've mentioned, the mathematical centerpiece of Part Four—Conrad's discussion of the converse of Hensel's Lemma—is developed specifically for whole p-adic numbers (no p-adic decimals). Mathematically stated, Conrad's discussion assumes that the p-adic norms of both the basic function f, and its derivative f', are less than or equal to 1.

This correspondence (that p-adic whole number means $|\ |_p \le 1$) makes sense because the definition of the p-adic norm is p^{-n} : If there are p-adic digits to the right of the decimal point, then n = -1 or -2 or -3 and so on, so $p^{-(-1)}$ or $p^{-(-2)}$ or $p^{-(-3)}$ is greater than 1, and is therefore excluded from the p-adic region discussed in Conrad's derivation of the converse of Hensel's Lemma.

This is no big deal for the serious mathematician, whose deep understanding of p-adic mathematics allows immediate intuition that cuts through issues such as p-adic scale, or schemes of p-adic normalization or renormalization.

However, our own intuitive development in Parts One through Three of *Digital Mind Math* often appealed to explanations using p-adic decimals (p-adic norm greater than 1). This is because we were often describing the Digital Mind Math of delving down into additional levels of detail, which can be intuitively modeled as adding more digits to the right of the decimal point.

But in Part Four—developed for p-adic whole numbers—our intuitive modeling, in order to allow room for more levels of detail (digits to the right), will often have a starting point with trailing zeroes: 123000 or 123000000 or 123000000000. This change in scale keeps us consistent with the assumptions we're using for the converse of Hensel's Lemma, but still allows for modeling of additional levels of cognitive detail.

Everyday Examples of the Converse of Hensel's Lemma

Following are examples that illustrate our everyday tendency to live our life according to the converse of Hensel's Lemma. These examples show our natural inclination to address issues that face us in the manner that the converse of Hensel's Lemma predicts is our evolved efficient way—that is, according to a comfortable, naturally appealing approach that makes incremental progress, or takes orderly steps, or builds up toward a solution, or inches toward a resolution, or refines an understanding, by:

- Finding an optimal pace of approach. The mathematical specification for this pace of approach is that our pace must be at summarization level M or lower (that is, level M or more detailed than level M), when faced with a $2M+1$ issue), and/or

- Lowering the number p of details examined at one time, so that the universal-p calculation of the p-adic magnitude $| f'(a_0) |_p$ of the derivative exceeds the square root $\sqrt{| f(a_0) |_p}$ of the magnitude of the issue at hand.

Unlike the mathematical specifications above, these examples will not be rigorously mathematicized. The intention is to convey how our mind, as its ordinary way of functioning, tends to push itself, yet pace itself, to optimally increase information content in accordance with the converse of Hensel's Lemma.

In other words, in the examples below, watch for cognitive patterns in which we conform to the converse of Hensel's Lemma —

We proceed at an optimal cognitive pace when we advance informational content:
- **By addressing component issues—issues with narrower scope, at a greater and more specific level of detail, and/or**
- **By selecting a smaller number of details to address at one time.**

Example 1. Moderate novelty and the zone of proximal development. We know from the work of Piaget that children learn new facts or information when they're ready for it, specifically when the new fact or information is "moderately novel." Vygotsky labels the region of moderate novelty the "zone of proximal development."

Of course, a lot of information is captured within the concepts of "moderate" and "proximal." This region depends on the information that the child has been previously exposed to, the child's capabilities with respect to how much new information can be absorbed at once, the inherent natures of the existing informational base and the new information, and so on.

So even a plain-English type of delineation is difficult and complicated for exactly what we mean by the ideal region of learning being a region of moderate novelty or a zone of proximal development.

Realistically, we cannot expect Digital Mind Math to leap over all of the psychological, pedagogical, and epistemological ambiguities in order to mathematicize both the child's current state of knowledge $f(a_0)$ and her region of moderate novelty or zone of proximal development (defined by the derivative $f'(a_0)$).

So we will have to settle, in Part Four, for advancing the science of Digital Mind Math somewhat by relating at least in a general or intuitive way Piaget's moderate novelty and Vygotsky's zone of proximal development to Conrad's converse of Hensel's Lemma.

With this in mind, here's the concept:

The child comes to a new piece of information with an informational structure that can be p-adically stated as a de Rham-Witt complex of sheaves of universal Witt vectors. According to the converse of Hensel's Lemma, the region with size $\mid f'(a_0) \mid_p$ of moderate novelty, also known as the zone of proximal development, is a region of information that is more specific (smaller in a real sense, which means that it is larger in a p-adic sense) than the entire region $\mid f(a_0) \mid_p$ of information that the child brings into this encounter.

The exact size of the region of moderate novelty, or the zone of proximal development, is defined as p-adically larger than $\sqrt{\mid f(a_0) \mid_p}$, which means that in a real sense it is within (smaller than; at a more detailed than; with a lower number of components than) the informational disk with this p-adic magnitude.

If we were operating in a simple p-typical p-adic environment (rather than a universal de Rham-Witt complex sheaves environment), this region is determined by the level of informational understanding that the child brings to the encounter: If the child's scope of understanding extends to 7 levels of complexity—if the child has a comfortable mastery of thinking about and discussing the topic with this real extent of scope—then the zone of proximal development is a region of more detailed concepts, not as general in scope: 3 or fewer levels of complexity. Optimal learning occurs with concepts this much less complex—level 3 or less—concepts with more detail and narrower focus, less of a big picture or large scope.

If the complexity of understanding that the child is bringing to the task extends to 2M+1 levels of detail, then the zone of proximal development is a region with M or fewer levels of detail.

This 2M+1 and M approach applies for the p-adically simpler p-typical environment, in which each level of detail has the same number p of details. In this uniform-p (p-typical) environment, moderate novelty, or the zone of proximal development, requires presentation of new information with a narrower scope.

There is an alternative, though, for which the new information presented to the child does not require such a narrowing of scope, such a delving down into more detailed levels of exposition. This alternative relies on a lower p, a lower number of details. In this universal-p (not p-typical) mathematical environment, we need to compute the minimum p-adic magnitude of the derivative by looking to the square root of the magnitude of the starting informational base. A sufficient lowering of p, a decrease in the amount of detail, will permit a concept of broad scope to still fall within the zone of proximal development. This would permit an entirely big-picture pedagogical approach, but the big picture would have to be described in summary terms, piece by piece, simplified, without much detail.

So the pedagogically optimal region of moderate novelty—the zone of proximal development—is a region described with M or fewer zeroes, and/or with a low enough value of p, so that $| f'(a_0) |_p > \sqrt{| f(a_0) |_p}$.

This seems pretty obvious—expose the child to a moderate depth of detail (rather than to the whole big picture at once), and/or to just some not all of the details at once—but no one knows how to mathematicize this precisely. We're fine with the idea that this seems pretty obvious. That's just the point—the converse of Hensel's Lemma has intuitive appeal. Our goal here is to provide an early mathematical framework for a very complex question.

Let's proceed with some more examples from different points of view to see how we can continue to proceed toward an intuitive understanding.

Example 2. Broken in just the right way. Comedian Bob Odenkirk—who has television roles on "Breaking Bad" and as star of "Better Call Saul," among other roles—was recently interviewed about how he sees himself as an actor and comedian. The interview turned to a question of how comedians can translate personal quirks or psychological issues into a successful comedy routine:

> "There are a couple of things wrong with me; some of them I make money off of. Ultimately what we're all doing is trying to turn our psychological problems into a paycheck. You want to be broken in just the right way to make the most amount of money."[91]

Here let's assume that Odenkirk is making money as a comedian based on the increase in informational content that he offers his audience by illuminating elements of their lives in a moderately novel way. Odenkirk's illuminatory approach involves funny spins on psychological issues that he personally faces, ways in which he is "broken."

Now this can't be a real downer of a psychological problem. And it can't be too small of an issue, either. Or one so specific to Odenkirk that it's not accessible to his audience.

Odenkirk has a lot to consider in order to draw from his personal psychological issues and define the boundaries of what makes a good joke. It must be a joke about being broken in just the right way.

In Digital Mind Math terms, Odenkirk is giving us the prose label for the region defined by the converse of Hensel's Lemma.

<u>Example 3. Captain Kirk refines Commander Spock's analysis.</u> In the 2013 film *Star Trek Into Darkness*, the forceful and passionate Captain James T. Kirk of the Starship Enterprise, and his ever-logical half-Vulcan Commander Spock, face two villains at once, creating a seemingly hopeless situation. John Harrison, formerly with Starfleet but now a terrorist, is actually the centuries-old evil superhuman Khan. And Kirk's commanding officer, Admiral Alexander Marcus, has his own evil secret agenda: sacrificing the Enteprise to an attack by the Klingon Empire in order to provide Marcus with the pretense for all-out war against the Klingons.

Kirk and Spock face many crises. At one point, Spock sets out the logical alternatives, constrained by military capabilities and protocol, twenty-third-century ethics, and the Prime Directive. But Spock's alternatives seem to Kirk tantamount to surrender. "I have no idea what I'm supposed to do," Kirk says. "I only know what I can do."[92]

Here we can assume that Spock has used his great Vulcan intellect to clarify the informational space of possible solutions that are available as the next step.

Spock is logical to his core, and it is certain that he has correctly calculated the region of all possible next steps that can occur subsequent to the current starting conditions.

Kirk does not doubt Spock's analysis: Spock has provided the universe of next steps that may logically proceed from the current step.

Kirk recognizes Spock's logic as providing all the possible next steps, but not specifically the best next step.

The optimal information-maximizing next step, Kirk knows, does not lie all the way to the edge of the configuration space created by circumstance, protocol, ethics, and the Prime Directive. Information maximization will best proceed by an approach that is more targeted, more precise, at more specific levels of detail (lower n) and/or fewer component details (lower p).

Kirk's focus is narrower than the informational region of what he is supposed to do, and is limited to the sphere of what he can do. This limit is the limit that—in this situation, in these circumstances—is defined for Kirk by the converse of Hensel's Lemma.

As a result, the plots of both Khan and Marcus are stymied, and (after a year's repair of the Enterprise) Kirk, Spock, and the team are sent on a five-year mission to boldly go where no one has gone before. What could create greater information maximization than that? How could anyone proceed at a more optimal pace of information maximization?

Example 4. What you want and what you need. The Rolling Stones tell us: "You can't always get what you want. But if you try sometime, you just might find, you get what you need."[93]

Isn't it clear that Mick Jagger and team are telling us that, as we examine our equation $f(x)$ of what we want, and we assess what we have $f(a_0)$ right now when $x=a_0$, the best incremental step $f'(a_0)$ right now is to focus on the region of what we need, which is a region that has some but not all of the complexity of our wants, that gives us a few—just what we need— but not all of the items on our wish list?

Example 5. More rock and roll. The musicians Daryl Hall and John Oates tell us:

> "I can go for being twice as nice. . . Now you want my soul. Ooh, forget about it. Now say, no go. . . I can't go for that. . . No can do. I can't go for that. Can't go for that. Can't go for that. . ."[94]

Here, once again, we can see members of the Rock and Roll Hall of Fame at their best, illustrating the converse of Hensel's Lemma.

In the case of Hall and Oates, we see the line drawn at being twice as nice: Mr. Hall and Mr. Oates can go for being twice as nice, so the p-adic magnitude p^{-n} of being twice as nice therefore has an exponent of −M or a smaller negative number, when faced with a −(2M+1) relationship arrangement.

But apparently being twice as nice is right at the edge of satisfying the conditions of the converse of Hensel's Lemma, since just a small amount more—my soul—when added to being twice as nice creates a condition with too high a p to still meet the conditions of the converse of Hensel's Lemma.

Being twice as nice creates a condition under which $| f'(a_0) |_p > \sqrt{| f(a_0) |_p}$. But just the small increment to p in the form of my soul—remembering that in the converse of Hensel's Lemma region of p-adic integers only, higher p means lower p-adic norm—creates a derivative with p-adic norm of magnitude less that the standard set by the converse of Hensel's Lemma.

Consequently: I can go for being twice as nice.

But twice as nice + my soul = No can do. No go. I can't go for that. Outside the optimal region set by the converse of Hensel's Lemma.

Example 6. Prequels. Many popular movies have sequels. And some have prequels, too—movies that take place before the original movie, that provide a back story for the characters and plot of the original movie.

For example, in the *Star Wars* series of movies, Episodes IV, then V, then VI were released, followed by Episodes I, II, and III. In the prequels, the writers did not have complete freedom to construct an informational space with all possible plot lines and character development. They had to bring us to where Part IV began. And to be really good at their jobs, they had to lead us to nod in agreement as they helped us complete our understanding of the situations and motivations of the characters as they proceeded beyond Part IV.

So the facts as already presented in Parts IV through VI constrained the possibilities for Parts I through III to a narrower informational space than the informational space that would have been available had the *Star Wars* series been released starting with Episode I.

Certain characters had to be alive, others dead, at the end of Part III. Some had to be evil, others good. Motivations and personality traits needed to be established. Galactic alliances needed to be established and explained. Whole planets needed to be in existence and settled.

In other words, the informational space of possibilities for Episode I of *Star Wars* would have been larger if Episode I was the first episode released than the informational space of possibilities for Episode I actually was, released after Episodes IV, V, and VI. Therefore, the informational space of possibilities for Episode I was constrained as the converse of Hensel's Lemma calculates to a smaller region of informational space.

Example 7. What makes the best plot twist? In Chapter 27 of Part Three, we examined the appeal of movies' plot twists as deriving from the natural pleasure that our biology gives us from increasing information content. But we mentioned that we'd be postponing until Part Four the discussion of how a plot twist can optimize the increase in information content. We are now ready to discuss this.

Obviously, this depends on the individual's entire past history of thoughts—what experiences she's had with movies, and with plot twists, and with confusion. Someone who's read or viewed a lot of mysteries or thrillers or fantasies or science fiction is going to require more complexity than a novice would. And there are also standards— which relate only partially to the individual, but which also have absolute components not dependent on the individual—for understandability, credibility, and enjoyability.

Of all possible plot twists, some aren't very good at all, and others depend on the individual viewer's or reader's circumstances. From all possible plot twists, the optimal plot twist is the one in the informational space defined by the converse of Hensel's Lemma.

Example 8. The well-designed book index. For many writers looking to ensure that their work is publication-ready *The Chicago Manual of Style* is the bible. It contains information on grammar, punctuation, usage, spelling, documentation, indexing, style, publishing processes, and so on.[95] How on earth can a useful index be created for almost 1000 pages crammed with this specific information?

For example, I needed to see, in writing the first line of this example, whether I should say "ensure" or "insure." My experience with *The Chicago Manual of Style* told me that I could use its index to rapidly converge on the answer, and I was right.

I took a chance on looking for "ensure" in the index, but it wasn't there. This didn't surprise or disappoint me, since I knew it was a long shot: Could every possible confusing word be listed separately in the index, or would that make the index so long as to be unusable?

What to try next? Well, my thought space around this ensure/insure question involved confusing words. "Confusing" didn't seem promising as an index term, so i went to "word" in the index, where I found "words" with some subentries under it, and "word usage," also with subentries—a total of 20 subentries between the two entries. A quick glance through these 20 subentries finds the location for "troublesome words," a 23-page alphabetically organized glossary of troublesome expressions (no wonder "ensure" itself wasn't in the index!).[96] There the "ensure, insure, assure" entry clarifies that *"ensure* is the general term. . . *insure* is reserved for underwriting financial risk. . . we *assure* people that their concerns are being met."[97]

What a brilliantly organized index!

I had no expectation that my first attempt—finding "ensure" itself in the index—would work. An index that contained so many details that it would include every troublesome word would be an index with too many details, too high a p, and therefore (in the region of the converse of Hensel's Lemma—p-adic whole numbers only—where high p means lower p-adic magnitude) not likely to produce a derivative—an indexing of the entire informational content—that is great enough (that is, greater than the square root of the informational content of the whole book).

But just the thought process that I went through as I looked up "ensure" led me right away to the "word" listings where I found "troublesome words" after a quick glance.

This for me was a remarkably efficient experience of the converse of Hensel's Lemma, much more efficient than a seriatim review of the entire table of contents, and a completely satisfying convergence at the index entry.

What did *The Chicago Manual of Style* do to make this such a satisfying converse of Hensel's Lemma experience? How did *The*

Chicago Manual of Style make use of the converse of Hensel's Lemma for its well-designed index?

To answer this, what better place would there be to look than the entry for "indexes" in *The Chicago Manual of Style* itself? Here we find 47 pages of advice, regarding general principles, subentries, cross-references, choosing terms, terms that should not be indexed, alphabetizing, and mechanics.[98] Since a useful index would not result from indexing every noun, verb, and adjective, we are advised to resist the temptation to index items that are "passing references" or "scene-setting elements" or otherwise "not essential to the theme of the work."[99] But "if many readers of a publication would be likely to look for their own names in the index[, o]ccasional vanity entries are not forbidden."[100] And don't forget: Although computerized "indexing software can streamline the indexing process . . . human intervention is always required."[101]

Obviously, an index consisting of too many words—or every word in the book—would not make a good index. While this would allow a user to find every term of importance (in this sense, it would be a sufficient approach), it would also include words that aren't important for indexing (it would not be a necessary approach). This is exactly when we need the converse of Hensel's Lemma: When we've found an approach that achieves the desired outcome but does so inefficiently—too many words, too broad or general, too large in a real sense, too small in a p-adic sense, of p-adic magnitude too close to the p-adic magnitude $p^{-(2M+1)}$ of the book as a whole—what we need is to examine the problem as the converse of Hensel's Lemma instructs us. We don't want to settle for a sufficient solution; we want the necessary solution. We need to examine the indexing process at a greater level of specificity and detail, to establish more specific rules and processes, to operate at a greater p-adic magnitude—specifically, at p-adic magnitude p^{-M} or greater.

Example 9. A great television news interview. What makes a great television interview with an important or influential public figure?

Personally, my favorite interviewer is Stephen Sackur of the BBC News interview program HARDtalk. Sackur interviews powerful world leaders one-on-one, and he comes prepared and is quite direct and forceful in his questioning and follow-up. He seeks illumination on the facts and on the justification for actions.

There is a certain baseline for what constitutes a reasonable interview, an acceptable interview. This would include coming prepared with a relevant list of questions, clearly stated. That's about all it takes to satisfy the minimum standards, to be sufficient for an interview. And, in my opinion, this is all you get from a typical television news interview.

What if you wanted to make it a great interview, one that maximizes the pace of transfer of informational content to the viewer?

To do this, you must subject yourself to more detailed standards. You need to not just clearly state a relevant series of questions. You must listen to the answers and ask follow-up questions. You must be prepared to a greater depth of detail so that you can respond intelligently. You must be grounded in an intellectual framework that permits you to deeply understand the context of the interviewee.

An ordinary interview increases information both locally and globally. But it is only an intensively focused and masterfully executed interview that optimally increases information content, because in an intense interview there is no fluff, no wasted time or wasted words or wasted effort.

An ordinary interview accomplishes what is sufficient for increasing information, but it does so in an inefficient way, by including elements that are not necessary for increasing information content.

Example 10. Is social media good or bad? A lot of people spend a lot of time on social media. Some of this involves cat videos and gossip about people famous for being famous. But we also entertained in Chapter 32 of Part Three the possibility that a higher purpose might be achieved by our involvement in social media. This higher purpose related to the creation of a universal human consciousness, one that would unite all earthlings with a shared universal intelligence, finding our place among other civilizations in the galaxy and universe.

To achieve this higher purpose, we cannot fall into lazy high-p social media habits of doing anything and everything that strikes our instant fancy. We must go into detail and organize the information, catalog the contents and the interconnections, understand how every electron and atom relates to every other electron and atom.

Clearly a job for the converse of Hensel's Lemma.

Example 11. I regret everything. In a 2015 Terri Gross NPR Fresh Air interview with Toni Morrison, celebrated author Morrison, now in her mid-80's, describes how she looks back at her life at times, and observes that "I remember every error, every word that I spoke that was wrong or incontinent . . . I remember everything as a mistake — and I regret everything."[102] She expands on this:

> "When I'm not creating or focusing on something I can imagine or invent, I think I go back over my life—I don't recommend this, by the way—and you pick up, 'Oh, what did you do that for? Why didn't you understand this?' Not just with children, as a parent, but with other people, with friends. . . It's not

profound regret; it's just a wiping up of tiny little messes that you didn't recognize as mess when they were going on."[103]

What a spectacular evocation of the converse of Hensel's Lemma!

Morrison poignantly verbalizes a thought process we go through under the influence of our drive for negentropy maximization, our drive to increase information content. This drive causes us, on occasion, to look back at mistakes we've made, and to try to envision what we should have said or done differently, how we could have fixed things, perhaps how we could still fix things. This is a direct implementation of Hensel's Lemma, where we know that if we fix things locally, it's guaranteed to fix things globally.

But Morrison takes this further, when she says "it's not profound regret; it's just a wiping up of tiny little messes." Here is where she uses the converse of Hensel's Lemma, to focus with eyes wide open at the detailed memory of an event that, decades later, she still recognizes as unresolved. The converse of Hensel's Lemma has brought her mind to this detailed focus—even at a situation of regret that is not profound, even at a tiny little mess—so that she can work at wiping up this tiny little mess before she goes back to bigger-picture creating and imagining and inventing.

Perhaps what Morrison says here resonates with you. Perhaps you recognize as a human urge—as your urge—this desire, this drive, to clean up tiny little historical messes. This drive results from our built-in, evolutionarily selected cognitive process that optimizes our information maximization process by addressing component issues, issues with narrower scope, issues with greater level of specific detail, issues that have a level of specificity smaller than the big broad issues that face us long-term.

And perhaps we can see why this small-bore focus can be psychologically healthy, can be an optimizing element of our lifetime information maximization.

Perhaps we can take advantage of Toni Morrison's intimate sharing of this aspect of how her mind works to use this as guidance if we're feeling a little self-critical about mistakes we might have made in relationships that are important to us: The need to clean up some little messes does not negate the good person that we are. Maybe we're not perfect, but they're still lucky to have us.

Example 12. Apologizing to people you've hurt in the past. To help with problems with alcohol, the organization Alcoholics Anonymous offers a Twelve Step program, which includes identifying people whom we have harmed and how we might make amends to them.[104] Why is this identification of past harms and how we can make amends relevant to helping address problems with alcohol?

This seems to be the identical issue that Toni Morrison identifies when she says, "I regret everything," and sets out to consider how historical personal messes can be cleaned up.

Within the Twelve Steps of AA. this is not hitting directly at the broad issue. Instead, it is a recognition of how addressing pieces of this broad issue—perhaps indirect pieces, several levels of detail down—will optimally contribute to the path toward a fuller resolution.

Example 13. Responding to racist or sexist comments or actions. One option, when subjected to a situation in which a racist or sexist comment is made in our presence, is to address the issue directly, to directly explain to the person who made the comment that you are offended and why. When would such a direct response be effective, and when would it be ineffective?

If the offensive comment was made carelessly, without much thought or consideration, and if it is not an integral aspect of an entire belief system, then a clear and direct response is more likely to be effective. For the offending commenter, the sentiment is not a large part of their whole personality, but rather an isolated comment. So 2M+1 is not a large number, freeing us up to respond more directly, with M not too deep or hidden or subtle. Perhaps 2M+1 is 3, so M can be 1, an information-expanding comment, but not too information-expanding, because not much is needed.

And if, as is likely, the offending commenter's mind is more complex that just p-typical—if his mind is organized as a de Rham-Witt complex of sheaves of universal Witt vectors—then we can still identify a response with informational content large enough to satisfy the converse of Hensel's Lemma. We can't find this informational place via the simplified 2M+1 and M route that is available only for p-typical situations; we have to instead find a derivative $f'(a_0)$ of the current informational state $f(a_0)$, this derivative having magnitude $| f'(a_0) |_p$ that is greater than $\sqrt{| f(a_0) |_p}$, the square root of the magnitude of the current informational state. We do this by selecting a limited number of specific details (low p) to comment on, not by elaborating about all of the ways (the many high-p ways) that this offensive comment matches so many other high-p aspects of their defective personality. By keeping p low, in the analytic framework of p-adic whole numbers not decimals, we can achieve a derivative $f'(a_0)$ with magnitude greater than the square root of the magnitude of $f(a_0)$.

But if the offensive comment was integral to the whole personality, reflecting a belief system seeping widely into all aspects of their thinking and acting, then it is unlikely that a straightforward retort would be effective. In all likelihood, all straightforward retorts have already been expressed by others previously, and a defensive wall of counter-retorts has already been built. 2M+1 is a big number, and we'll need to go many levels of detail down—subtler, not straightforward— if we want to create a meeting of minds. This big project will have to start very small.

Or maybe we judge that it would be too hard to do this, or just not worth the effort, or perhaps even that we're facing a dangerous situation. So we walk away. This does not resolve the issue (increase aggregate informational content); it just opens up a new unrelated informational space from which we take our next step in life.

Example 14. Why is revenge best served cold? There is an expression that revenge is best served cold. This means that sometimes we obtain more satisfaction by not responding directly, but rather by biding our time, waiting for a more opportune moment to exact our revenge. Why is this?

Perhaps this is because the only responses available to us immediately are too similar to the event that caused us harm. Maybe the harmful event had a level 7 of detail, and all we could muster up as a response was something that responded pretty directly—at a level 6 of detail—which is too close to the situation that we had already been beaten at. A level-6 response to a level-7 event does not satisfy the conditions of the converse of Hensel's Lemma.

Give it time. Let the situation unfold. What our enemy stole from us will become embedded into his life. And we'll have time to contemplate and to prepare in detail.

Later on—when the situation has appeared to cool off—you can bet that we'll be prepared with a subtle jab that will give us the satisfaction of our revenge!

Example 15. Walking on eggshells. Have you ever had a fight with a friend or spouse? Maybe for a while, after the fight, we interact more gingerly, cautiously, tentatively, tenderly. We walk as if on eggshells.

We keep our psychological and interpersonal distance from each other, as we search for a more distant level of details from which we can begin to repair the the relationship.

How to implement the converse of Hensel's Lemma is not always immediately obvious. Rather than risk suboptimization with a response too close to the conflict, maybe we should give ourselves a little time and space to consider what this is all about.

Example 16. Why something bothers (rather than just amuses) you. Have you ever been annoyed by something or someone that just seems to amuse other people? Or maybe you've wondered why something bothers other people, when you just find it funny?

Often something or someone bothersome is too close for comfort, too close to our own unresolved issues for us to be at peace with. We cannot reach equilibrium with this situation because the issues it presents us with are not as informationally distant as the converse of Hensel's Lemma requires. The situation threatens us.

To find humor in a situation, we need a certain psychological distance. The converse of Hensel's Lemma gives us guidance as to where this is.

Example 17. Why want anything more marvelous than what is? Longtime book editor and essayist Diana Athill ends her essays and memoir Alive, Alive Oh!, written at age 98, with a poem including the lines:

"Why want anything more marvelous
than what is"[105]

Can there be a more eloquent way to end our discussion of the converse of Hensel's Lemma?

We want a lot. The informational space of what we want is large, but we still want more. Will we ever be at peace with all that we want? Where is the informational space of equilibration?

Look at who and where you are, at what you have, at who and what is around you. Look at the whole picture and at every detail.

According to the converse of Hensel's Lemma, you will best optimize your information content when what more you want—the derivative of what you now have—has a p-adic magnitude that is greater than the square root of the p-adic magnitude of what you now have. You will achieve this by focusing on a handful of details within a small component of your life—by appreciating the beauty around you, by enjoying life's simple pleasures, by stopping to smell the roses.

What Part Four Says About Digital Mind Math

The hope is that this dance with the converse of Hensel's Lemma, these brief everyday examples many levels of detail more specific than the general mathematical statement that $| f'(a_0) |_p$ must exceed $\sqrt{| f(a_0) |_p}$, will help form a conceptual image beyond the earlier Parts' mathematics of the organization and processes of the mind, all the way to mind optimization.

Conclusion: Minds, Brains, and Computers

Let's conclude with some speculation.

This speculation will have as its theme the application of the Digital Mind Math of this book not only to the mind, but also of two related fields of inquiry—the brain, and artificial intelligence (computers).

Thus, in this Conclusion, we take an Occam's Razor approach—an approach assuming that the simplest solution is the best solution. In particular, we explore the implications of applying the mathematics of Digital Mind Math to the science of the brain and the science of artificial intelligence, based on Occam's Razor assumptions that (1) the brain and the mind are organized and structured similarly, and (2) the mind's (and the brain's) organization and structure are the best model for artificial intelligence.

We'll warm up for this discussion with a brief digression to the philosophy of mind, and a conjecture as to the nature of consciousness, thought, memories, and all that we experience as the mind.

What Is the Mind?

What we experience as the mind is our sensation of the brain.

When our finger touches a hot stove, we experience a certain set of feelings. When light reaches our eyes, or sound waves reach our ears, we have conscious experiences of what is going on around us, what we're seeing and hearing, and what it all means.

When we think, we are experiencing the electromagnetic and chemical interactions that are taking place within our brain.

What an exquisite evolutionary advancement thinking is!

Well beyond simple automatic responses to simple external stimuli, we have evolved to respond to complex aggregations of electrochemical stimuli received by signal-receiving neurons within our brain, passed across synapses from signal-emitting neurons.

The mind is our name for the sensation we feel as a result of the electrochemical activity of the brain.

Neuroscience

According to Digital Mind Math, our natural cognitive mathematics is the incredibly powerful mathematics of enclosure, p-adic mathematics.

P-adic numbers label the paths of signals across a succession of neurons. Because of the computational efficiency that p-adic Witt vectors allow, whole sets of paths can be examined and manipulated at once.

According to Digital Mind Math, there is a three-step process to thinking: First we generate possible next thoughts. Then we select an information-maximizing thought. Then we experience the thought that was selected.

The brain has evolved to excel at this three-part process. The possible next thoughts are every combination, starting from the previously experienced thought, of sequences of neurons emitting signals to sequences of receiving neurons. Information maximization results from longer chains of neurons, and from neurons connecting to more and more sets of other neurons. Neurobiological processes, evolved based on p-adic mathematics, identify and experience neuronal sequences that maximize information by lengthening chains and increasing their connections—that is, sequences that maximize negative entropy.

In Dr. Matti Pitkänen's documentation of Topological Geometrodynamics, you will find extensive elaboration of biophysical evidence and mechanisms corresponding to the Digital Mind Math that this book summarizes. To whet your appetite, here are some examples of concepts from TGD biophysics:

- Biosystems as superconductors
- Quantum antennas
- Massless extremals
- Information molecules
- The four-dimensional brain
- Biosystems as conscious holograms
- Topological quantum computation
- DNA as topological quantum computer
- Hologram-generating properties of DNA
- Genetic code and the role of dark matter in biosystems
- Quantum model of the electroencephalogram (EEG)
- TGD-inspired model of nerve pulses
- Hyper-genome coding for cultural and social evolution
- Magnetospheric consciousness

Dr. Pitkänen has generously presented an introductory overview of some of these mechanisms in his Afterword included in this book. However, true justice to the depth and breadth of Pitkänen's work in this regard can be achieved only through study of Pitkänen's full original work, which he has made easily and freely available online. As of this writing, the best source for this is Dr. Pitkänen's TGD website www.tgdtheory.fi. Dr. Pitkänen also regularly publishes musings on physics and consciousness at www.matpitka.blogspot.com. Or feel free just to browse by searching using the terms "tgd pitkanen."

Artificial Intelligence

It appears that, of the two complete number systems, computer science has picked the wrong one. Computer science will proceed better using p-adic mathematics than it has using real mathematics. P-adic mathematics offers an efficiency that is orders of magnitude superior to approaches based on real binary mathematics.

Computer science's great successes today are overly reliant on the brute-force advantages of high processing speed and high data volume capability. As a result, todays's computer science applications require industrial-size processing, power, and cooling systems.

It is time for computer science to focus on the efficiency of the three-pound twenty-watt human brain. P-adic mathematics, Digital Mind Math, and Topological Geometrodynamics show the way.

Current computer science theorizing seems on the verge of finding p-adic mathematics as the natural mathematics of artificial intelligence. Approaches such as fuzzy technology, soft computing, and possibility theory seem ripe for exploration with p-adic mathematics rather than real binary mathematics.

Perhaps the most radical possible applications of p-adic mathematics to computer science relate to the basic infrastructure of computer hardware.

How could this be accomplished? Using biotechnology (biology-based computer hardware)? Using quantum computer methodologies, based on quantum phenomena such as quantum entanglement or quantum superposition? Using engineered smart biomaterials, or optical metamaterials,[106] or organometallic polymerization[107]?

Relationship to Current Scientific Research

If you enjoy keeping up with contemporary research efforts in the fields of cognition, neuroscience, or artificial intelligence, you will likely come across current scientific research that seems to fit well within the Digital Mind Math framework.

Three broad aspects of the Digital Mind Math framework which seem to also recur as significant themes in these fields of contemporary scientific research are:

- Bounded concepts, intricately linked
- Cognitive science inspired by quantum physics
- Biology-based artificial intelligence

Here are some technological developments--some at a purely theoretical stage, some at early stages of application—that may stimulate your imagination for how Digital Mind Math could be applied:

<u>Bounded concepts, intricately linked</u>—the organization of the mind, consistent with the $M^4_+ \times CP_2$ structure of Topological Geometrodynamics physics, and also captured with extraordinary efficiency by the structure offered by p-adic mathematics:

- Mining patterns from databases[108]
- Algorithms for identifying and evaluating subspace clustering of data streams[109]
- Multi-granular computing to optimize intelligent processing of big data.[110]
- Granularity clustering—uncertain artificial analysis combining granular computing with clustering analysis[111]
- Kernel information propagation tag clustering algorithm—using density information in the kernel neighborhood to cluster information[112]
- Condensed representations and interestingness measures for data set mining[113]
- Constructing and revealing "phantasms"—cognitive phenomena and ideas identified within data structures and computational media[114]
- Random sets and fuzzy sets as ill-perceived random variables (higher-order uncertainty)[115]

- The linear complexity profile and correlation measure of interleaved sequences[116]—remarkably resonant with TGD's and Digital Mind Math's $M^4_+ \times CP_2$ structure (interleaved sequences) and quantification of information by p (linear complexity) and negative-n (correlation)
- Unconscious-conscious interconnection modeled with information trees and neuronal trees[117]—Resonant with Digital Mind Math's dual modeling with p-adic mathematics (unconscious) and real and complex mathematics (conscious).Note that, although this work refers to p-adic models, it expresses a preference for m-adic models, where m is any whole number, not just prime numbers. The Digital Mind Math approach would here take advantage of Universal Witt Vectors, which offer the mathematical advantages of p-adic numbers and also permit non-homogeneous information trees.
- Horizontal and vertical correlation cryptograpic attacks[118]—Models of these "side-channel" threats to cryptographic systems resonate with the high-p (horizontal) vs. high-negative-n (vertical) alternatives offered in Digital Mind Math's generation of possible next thoughts.
- Dimensionality reduction techniques applied to the high-dimensional, complex geometry of the human brain connectome[119]
- Hierarchical pattern self-assembly based on algorithm using DNA origami tiles[120]
- Ultrametricity in complex systems[121]
- Identifying the fractal and oscillatory components of neurophysiological signals by irregular-sampling auto-spectral analysis[122]
- Hyper-lattice model for data systems[123]
- Achieving robotic pattern recognition and contextual

awareness by integrating video, ultrasound, radar, sonar, laser, and infrared information, in imitation of the way that "the whole visual system shreds images, breaks them into maps of colour, maps of motion, and so on, and somehow then manages to reintegrate that"[124]

- Computational intelligence (CI)—this field, developed over the last two decades, draws from the fields of neural networks, evolutionary computing, and fuzzy logic[125]

Cognitive science inspired by quantum physics—applying quantum physics to the mind, as Digital Mind Math does by virtue of its core cognitive process being the same process as quantum physics' core process, the quantum jump, and as Topological Geometrodynamics does with its parallel development of cognitive science, particle physics, cosmology, and biophysics:

- Cognitive decision-making as the collapse of a quantum superstate—new model, based on quantum random walk theory, in contrast to Markov random walks[126]
- Using quantum theory formalism to explain concept combinations and human reasoning as the superposition of conceptual reasoning (emergence of new conceptuality) and logical reasoning (algebraic calculus)[127]
- Quantum simulators[128]
- Data compression of qubits (quantum bits)[129]
- Quantum spin control via dark states[130]
- Holographic memory devices exploiting spin wave interference[131]
- Repeatedly blocking a quantum system's coherent evolution through measurement back-action (the quantum Zeno effect)[132]
- "Twisting" light to encode separate channels of

information[133]

- Examining the behavior of physical fields in "mesoscopic spacetime"—a possible intermediate spacetime regime, between the microscopic domain of quantum particle physics and the macroscopic scales of classical physics and general relativity[134]

- Measurements of identity and conceptual indistinguishability in human thought show Bose-Einstein statistics consistent with the presence of microscopic quantum particles[135]

- A formal analysis of music composition, applying quantum computational logic to illuminate vague and ambiguous features of musical ideas and extra-musical meanings[136]

- The advantages of many-valued, rather than 2-valued, logic to help understand quantum phenomena[137]

- Hybrid protocols to quantum information processing, combining discrete qubits with continuous Gaussian states[138]

- Quantum theory is an information theory[139]

- Projects of the Quantum Computing and Devices group at the Academy of Finland Center of Excellence in Computational Nanoscience (Aalto University):
  - o silicon nanoelectronics
  - o monopoles in Bose-Einstein condensates
  - o superconducting[140]

Biology-based artificial intelligence—inspired, as Digital Mind Math and TGD are, by biological processes' success with cognitive functioning:

- Quantum Information Biology (QIB)—"a multi-scale model of information processing in bio-systems: from proteins and cells to cognitive and social systems"[141]

- Nanomodification of living organisms by biomimetic mineralization[142]
- Conjugation of DNA to colloidal quantum dots or semiconductor nanocrystals[143]
- Sensors that detect miniscule vibrations by copying the design of an organ found in spiders' legs[144]
- The molecular processes and rheological properties in the self-assembly of spiderweb nanofilament[145]
- Biologically inspired silicon-based materials[146]
- Exploiting light-sensitive proteins (opsins) to control the brain with colored light that instantaneously activates or silences neurons[147]
- Organizing and rearranging central pattern generators (CPGs)—pattern-generating neuron ensembles, originally studied in the context of motor activities, now being examined as possible mechanisms for control of non-motor brain functions[148]
- Object recognition in natural settings by segmenting unseen, artificially created stimuli derived from category information[149]
- *Design of Intelligent Systems Based on Fuzzy Logic, Neural Networks and Nature-Inspired Optimization*—a 47-chapter anthology of applications to intelligent control, robotics, pattern recognition, prediction, and optimization.[150]
- IBM's TrueNorth neuromorphic chips—programmable silicon "neurons" and "synapses"[151]
- DNA computers: DNA "walker" circuits, having natural interfaces with both chemical and biological systems[152]
- Computational DNA circuit algorithm inspired by ant foraging system[153]
- Nature-inspired computing (NIC)—this emerging field draws

> from both biology-based and physics-based approaches and
> algorithms[154]

Digital Mind Math is the operationalization of the aspects of Dr. Matti Pitkänen's Topological Geometrodynamics that relate to cognition and the mind. I hope this book helps to promote TGD as the basis for understanding the mind and the brain, and also as a twenty-first-century theory of modern physics.

Afterword, by Dr. Matti Pitkänen

I have enjoyed reading Mr. Paster's expansive thoughts on the Topological Geometrodynamics model of cognition and the mind. This book, *Digital Mind Math*, presents a broad and largely nontechnical development of this aspect of TGD, and it is a welcome supplement to my own writings on the subject.

This book is the second book by Robert Paster related to TGD. The previous book was about TGD as it was a decade ago. In his current book, Robert does an excellent job explaining in popular terms those aspects of TGD which relate to the understanding of cognition and imagination—or more precisely, their mathematical correlates—in terms of p-adic mathematics and physics. For a long time I thought that also intentionality could be understood p-adically but it is perhaps better to identify imagination as the p-adic aspect of intentionality having both real and p-adic aspects.

There is a quite impressive number of applications of TGD to consciousness and living matter. TGD has evolved strongly during the last decade, and this is of course not discussed in the first book by Robert. Hence my Afterword is a summary of the basic principles and ideas relevant for TGD inspired theory of consciousness, quantum biology, and cognition. The style is perhaps rather technical—I do not have the genes of a populariser—but I dare hope that Robert's representation provides background making it easler to understand what I am saying.

I hope the interested reader will be able to find the time to explore TGD at my website www.tgdtheory.fi , where I continue to update extensive documentation of TGD's models, not only of cognition and the mind, but also of particle physics, cosmology, and biophysics.

I present the following thoughts as an introduction:

1. Why do we perceive world as 4-dimensional?

Kaluza-Klein theories and super string models encourage to think that space-time could have dimension higher than 4: say 10 or 11. In fact, any attempt to explain the quantum numbers of particles requires the presence of higher-D space. Usually one thinks in Kaluza-Klein spirit that these additional dimensions are very small so that we cannot "see" them.

In TGD framework the explanation for the experienced 4-dimensionality is different. We perceive world as 4-dimensional because we ourselves are 4-dimensional! We correspond to particular 4-D space-time surfaces in 8-D space-time $H = M^4 \times CP_2$ imbedding space, which is Minkowski space M^4 with points replaced with extremely small 4-D space CP_2 with size of order 10^{-30} meters. Also the

8-D imbedding space H looks 4-D Minkowski space M^4 in the resolutions that we usually use but this would not explain why we experience world as 4-dimensional.

As a matter of fact, despite the fact that CP_2 is so small, we actually experience the CP_2 degrees of freedom indirectly: CP_2 explains part of standard model quantum numbers. Even more, visual colors could correspond to quantum numbers assignable to CP_2 degrees of freedom.

This picture leads to the notion of many-sheeted space-time for which a good 2-D illustration is in terms of slightly deformed parallel planar surfaces connected by small wormhole contacts.[155] The picture of wormhole can be found from every popular book of general relativity.[156] What is new and fascinating is that wormhole contacts have Euclidian signature of induced metric: all 4 directions are spatial, and there is no direction which could define time coordinate.

These wormholes differ in many respects from those of general relativity (no space-ships traveling through them!) although the topology is the same. In general relativity theory (GRT) the wormholes are not stable. Also in TGD this is true unless the wormhole contact carries magnetic monopole flux. The conservation of this flux prevents the splitting. Due to their extreme smallness we do not "see" these wormholes directly but elementary particles are made of them so that we "see" them in a more abstract sense.

The magnetic flux from wormhole throat travels along space-time sheet to another wormhole throat through it to another space-time sheet and returns to the opposite throat of the original wormhole contact so the closed magnetic flux loops are obtained: from A to B along first space-time sheet and from B back to A along second space-time sheet! The so called ER-EPR correspondence[157] proposed by Susskind and Maldacena in 2013 states that in General Relativity blackholes are connected by wormholes having similar function as flux tubes in TGD framework. The instability of wormholes is the basic objection against this idea. In TGD Universe the stable monopole flux tubes have been serving similar function for more than decade.

2. What are the thought bubbles made of?

Descartes introduced what is known as dualism. Two kinds of substances would exist: matter and thought. One can challenge the idea about thought as substance: thoughts are about something unlike matter, which just is. Thoughts might however have physical and even geometric correlates. We have developed fantastically precise theories for matter and to high extent geometrized physics of matter, but what the correlates for thought bubbles might be has remained an enigma. p-Adic physics provides a possible solution of this enigma.

2.1. Why p-adic numbers should relate to cognition

p-Adic number fields[158] labelled by primes $p=2,3,5,...$ provide an infinite hierarchy of number fields analogous to reals: they allow basic arithmetic (sum, multiply, subtract, divide) and are obtained as completions of rational numbers obtained by adding something to fill the gaps between discrete rationals in order to obtain a continuum. To be precise, rationals allow an infinite hierarchy of algebraic extensions obtained by adding algebraic numbers and for these completion to extensions of p-adic number fields makes sense too. A gigantic infinity of number fields is patiently waiting for their application to physics!

1. p-Adic numbers are not well-ordered and p-adic notions of distance and continuity differ
from their real counterparts

p-Adic notions of size and continuity differ from their real counterparts: most of p-adic numbers are infinite as real numbers since they correspond to infinite integers with strictly infinite number of positive digits in their expansion in increasing powers of p (analogous to decimal expansion). For instance, the number $1+p+p^2+...$ is infinite as a real integer but has p-adic norm equal to 1. The p-adic distance $d(x,y)$ is measured by the p-adic norm of $x-y$ and is $p^{(-N)}$ for x and $y=x+p^N$, and thus arbitrarily small for $N>0$ large enough. In real context p^N would have arbitrarily large size. Hence the p-adic notions of nearness and of continuity differ dramatically from their real counterparts. The nearness of x and $x+p^N$ implies that fractality is inherent for p-adic numbers.

p-Adic topology is also very rough: all points having same norm have same "absolute value"—or more technically norm. In the real context only the points x and $-x$ have the same absolute value. About points with same p-adic norm one cannot tell which is the larger one. One says that p-adic numbers are not well-ordered.

Finite measurement resolution is a natural physics counterpart for the lack of well-orderedness: with a finite measurement resolution it is not possible to say which of the two measurement outcomes within the resolution is larger. p-Adic physics is thus tailor made for describing finite measurement resolution. One can classify things by putting things not distinguishable by measurement resolution to the same class—bin. p-Adic norm realizes this mathematically. Cognition indeed always classifies: two things which do not differ too much are regarded as same.

This classification prevents us from drowning to irrelevant information. For instance, the lowest bit of 2-adic number $b_0 + b_1 \times 2 + b_2 \times 2^2 + ...b_0$ is the most significant one. The higher bits add only details and in infinite resolution the number of bits is infinite. Pinary expansion defines a path in tree in which each branch divides into p branches. This is also a manifestation of p-adic fractality. Also decimal expansion has this tree character but expansion in powers of 10 is by no means inherent to real numbers—any base for the expansion is possible; for p-adic numbers only single prime *p*

2. p-Adic differential calculus is non-deterministic and describes imagination

p-Adic number fields allow also differential calculus making possible to build physics based on differential equations. There is however a strange phenomenon involved. p-Adic integration constants, such as the initial position and velocity for a particle falling in gravitational field, are p-adic pseudo constants meaning piecewise constancy. They have vanishing p-adic derivative and this is enough. This implies non-determinism (non-predictability), naturally identifiable as the non-determinism of imagination. Imagination breaking the laws of real physics would correspond to the physics of p-adic world. Besides reality one would have infinite number of p-adicities, the worlds only imaginable!

This leads to the idea that also p-adic space-time surfaces in p-adic counterpart of 8-D imbedding space are possible and provide fundamental cognitive representations for the real physics accompanying even elementary particles. Cognition would not be something associated with only brain but a universal aspect of existence.

3. How to unify real and various p-adic physics together

One must somehow unify real physics of matter (of sensory experience) and p-adic physics of cognition for various primes *p* to single coherent structure. Here the basic observation is that all these number fields have rationals in common. Rational world would be common to both cognition and sensory experience. Until rather recently I thought that this could be realized at the level of space-time surfaces: rational points of space-time surface would represent the intersection of thought and matter.

My recent view is more abstract: the intersection of realities and p-adicities consists of classical worlds (space-time surfaces) rather than their points! The points in the world of classical worlds (WCW) with rational WCW coordinates are these common points: rational worlds, one might say. The conceptual leap from space-time surface as world to WCW formed by all space-time surfaces is huge but solves a lot of conceptual problems and leads to understanding of basic aspects of imagination and cognition. For instance, space-time discretization leads to problems with symmetries: at WCW level this problem is avoided.

Mathematicians have an elegant name for the fusion of various number fields along common rationals: adelic construction.[159] The Universe and physics are adelic. Matter and thought are unified in adelic physics and infinite hierarchy of intelligences is predicted. The magic thing is by its gigantic symmetries this adelic WCW can be handled mathematically.

2.2 Some examples about p-adic concepts related to biology and consciousness

The first application of p-adic physics was to elementary particle physics—more than thirty years ago. It turned out possible to calculate elementary particle masses in simple model based on p-adic thermodynamics.[160] Number theoretic constraints made the model extremely predictive and the predictions were excellent. I had however to assume that the preferred p-adic primes come as primes near power of 2: $p \cong 2^k$, k integer. They correspond to p-adic length scales $L(k) \propto 2^{k/2}$ coming as half octaves and characterizing Compton lengths of particles. Recently this hypothesis has found a deeper mathematical justification and has been also generalized so that the primes $p \cong q^k$, $q = 2,3,5,...$ and k integer, are favoured. Music would represent example about experience involving both $q = 2$ and $q = 3$.[161]

In the following some example about applications of p-adic physics to living matter systems.

1. Mersenne primes are the primes which are nearest to a power of 2: $M_n = 2^n - 1$ and expected to be very special physically. This turns out to be the case. There are not too many Mersenne primes and $M_{127} \cong 10^{38}$, which corresponds to electron is the largest Mersenne prime which does not correspond to completely super-astrophysical p-adic length scale. The corresponding p-adic time scale is .1 seconds, defining a fundamental biorhythm. Hardly an accident! For instance, alpha band of EEG often assigned to calm state of mind and creativity corresponds to 10 Hz frequency.

   This suggests that p-adic physics is fundamental in living matter and that p-adic cognition reflects itself also in the real physics although we cannot observe p-adic space-time sheets directly: measuring the weight or size of thought does not make sense.

2. Further encouragement comes from the observation that there are 4 Gaussian Mersennes (primes for complex integers $m + in$) $M_{G,k} = (1 + i)^n - 1$, $n = 151, 157, 163, 167$, correspond to four biologically fundamental length scales between cell membrane thickness and the size of cell nucleus.

3. The third application is the notion of negentropic entanglement (NE). NE makes sense if entanglement probabilities are rational or in algebraic extension of rationals. If so, one can modify the definition of entanglement entropy based on Shannon formula by replacing the logarithm of probability with the logarithm of p-adic norm of probability. The outcome satisfies the axioms for entropy but there is a surprise in store. This entropy can be negative and is smallest possible for a unique p-adic prime. The proper interpretation is as information about something rather than as lack of information about the state of either entangled particle. The conscious information would be about the entire entangled system and entanglement would give rise to this information. NE is assumed to play a fundamental role in living matter.

3. Strong form of holography and imagination

Holographic principle[162] has become one the basic pillars of theoretical physics during last decades. It says roughly that 3-dimensional data are enough to fix 4-dimensional physics. One would have two dual descriptions provided by 4-dimensional physics and 3-dimensional physics—both simple in some but in different aspects.

The so called AdS/CFT correspondence[163] discovered by Maldacena provides a possible mathematical realization of holography and is inspired by superstring models although it is not part of super string theory as such. In TGD the holography becomes dramatically simpler and also stronger principle thanks to generalization of conformal symmetry—basic symmetry of string models—from 2-D context to 4-D context.

In TGD holography generalizes to what I call strong form of holography (SH) and follows from an extension of General Coordinate Invariance—second key principle in General Relativity. It states that 2-dimensional data assignable to string world sheets and partonic 2-surfaces—"space-time genes" one might say—code for 4-dimensional space-time physics. This is very much like the ordinary holography but is dynamical.

SH is very powerful principle and simplifies enormously the physics of single space-time sheet. Many-sheeted spacetime is however engineered from a large number of space-time sheets and at general relativity limit of TGD replacing these sheets with single slightly curved region of Minkowski space the fundamental simplicity is lost. The challenge for experimentalists is to discover how to measure what happens at single space-time sheet.

One of the implications of SH is rather convincing justification for the identification of p-adic physics as physics of cognition and imagination. One can construct the space-time surface by "algebraically continuing" the 2-D string world sheets and partonic 2-surfaces to 4-D surfaces: the process is somewhat like the analytical continuation in complex analysis perhaps familiar to some readers. It allows to assign to real function a complex valued function by replacing real argument with a complex one. This procedure makes sense for both real sector (sensory world) and p-adic sector (cognition and imagination).

One can say that string world sheets and partonic 2-surfaces belong to the intersection of realities and p-adicities and are characterized by WCW coordinates, which are rational or in some algebraic extension of rationals. The continuation in real case is rather unique. In p-adic case the already mentioned possibility of p-adic pseudo-constants makes the continuation highly non-unique: this is nothing but the non-determinism of imagination. One can imagine all kinds of things but very few imaginations are realizable!

4. Hierarchy of Planck constants, dark matter, and living mater

By quantum classical correspondence space-time sheets can be identified as quantum coherence regions. The fact that they have all possible size scales strongly suggests that Planck constant must be quantized and can have arbitrarily large values so that coherence regions, whose size are typically proportional to Planck constant, can be arbitrarily large.

If one accepts this then also the idea about dark matter as a macroscopic quantum phase characterized by an arbitrarily large value of Planck constant emerges naturally as does also the interpretation for the long ranged classical electro-weak and color fields predicted by TGD. It is now clear that the hierarchy of Planck constants follows at deeper level from the basic symmetries of TGD. Also the quantum criticality of TGD implies it: quantum critical systems involve large h_{eff} phases and fluctuations between phases with different values of h_{eff}.[164]

4.1. Dark matter as phases of matter with large gravitational Planck constant

D. Da Rocha and Laurent Nottale[165] have proposed that Schrödinger equation with Planck constant $\hbar$ replaced with what might be called gravitational Planck constant $\hbar_{gr} = GmM/v_0$ ($\hbar = c = 1$). v_0 is a velocity parameter having the value $v_0 = 144.7 \pm .7$ km/s giving $v_0/c = 4.6 \times 10^{-4}$. This is rather near to the peak orbital velocity of stars in galactic halos. Also subharmonics and harmonics of v_0 seem to appear. The support for the hypothesis coming from empirical data is impressive.

Nottale and Da Rocha believe that their Schrödinger equation results from a fractal hydrodynamics. Many-sheeted space-time however suggests that astrophysical systems are at some levels of the hierarchy of space-time sheets macroscopic quantum systems. The space-time sheets in question would carry dark matter.

Nottale's hypothesis would predict a gigantic value of h_{gr} . Equivalence Principle and the independence of gravitational Compton length on mass m implies however that one can restrict the values of mass m to masses of microscopic objects so that h_{gr} would be much smaller. Large h_{gr} could provide a solution of the black hole collapse (IR catastrophe) problem encountered at the classical level. The resolution of the problem inspired by TGD inspired theory of living matter is that it is the dark matter at larger space-time sheets which is quantum coherent in the required time scale.[166]

It is natural to assign the values of Planck constants postulated by Nottale to the space-time sheets mediating gravitational interaction and identifiable as magnetic flux tubes (quanta) possibly carrying monopole flux and identifiable as remnants of cosmic string phase of primordial cosmology. The magnetic energy of these flux quanta would correspond to dark energy and magnetic tension would give rise to negative "pressure" forcing accelerated cosmological expansion. This leads to a rather detailed vision about the evolution of stars and galaxies identified as bubbles of ordinary and dark matter inside magnetic flux tubes identifiable as dark energy.

4.2 Hierarchy of Planck constants labelling phases of dark matter from the anomalies of neuroscience and biology

I ended up with the hierarchy of Planck constants from neuroscience, but also the many anomalies of biology such as bio-photons[167] support the view that dark matter might be a key player in living matter. As a matter of fact, biology as such represents a huge anomaly probably understable only if one accepts macroscopic quantum coherence impossible in standard quantum physics.

The quantal ELF effects of ELF em fields on vertebrate brain have been known since seventies. ELF em fields at frequencies identifiable as cyclotron frequencies in magnetic field, whose intensity is about 2/5 times that of Earth for biologically important ions have physiological effects and affect also behavior. What is intriguing that the effects are found only in vertebrates (to my best knowledge). The energies $E = hf$ (f denotes frequency) for the photons of ELF em fields are extremely low—about 10^{-8} times lower than thermal energy at physiological temperatures—so that quantal effects are impossible in the framework of standard quantum theory. The values of Planck constant would be in these situations large but not gigantic.

This inspired the hypothesis that ELF photons correspond to so large a value of Planck constant that their energy $E = h_{eff}f$ is above the thermal energy. The conjecture is that the spectrum of Planck constants is given as integer multiples of the ordinary Planck constant: $h_{eff} = n \times h$. This leads to a proposal for the identification of the mysterious dark matter[168] about which we only know that it exists!

The proposed interpretation of the large h_{eff} photons indeed was as dark photons. A more general hypothesis was that dark matter corresponds to ordinary matter but in phase with non-standard value of Planck constant. If only particles with the same value of Planck constant can appear in the same vertex of scattering diagram, the phases with different value of Planck constant are dark relative to each other. The phase transitions changing Planck constant can however make possible interactions between phases with different Planck constant but these interactions do not manifest themselves in particle physics. Also the interactions mediated by classical fields should be possible. Dark matter would not be so dark as we have used to believe.

The further hypothesis $h_{eff} = h_{gr}$—at least for microscopic particles—unifies the two conjectures and would mean that quantum gravitation would be fundamental for consciousness albeit in sense totally different from that in the conjecture of Penrose.[169] In particular, implies that cyclotron energies of charged particles do not depend on the mass of the particle and their spectrum is thus universal although corresponding frequencies depend on mass. In bio-applications this spectrum would correspond to the energy spectrum of bio-photons[170] assumed to result from dark photons by h_{eff} reducing phase transition and the energies of bio-photons would be in visible and UV range associated with the excitations of bio-molecules.

4.3. Hierarchy of Planck constants, dark matter, living matter, and quantum criticality

Biological systems are critical systems—we survive only in a finite temperature range of few degrees—and therefore highly sensitive to changes in the environment. Living systems are also rather unpredictable—just like systems at thermodynamical or quantum criticality. Note that quantum criticality[171] generalizes thermodynamical criticality: thermodynamical non-deterministic long range fluctuations are replaced by their quantum counterparts.

One of the basic assumptions of TGD is that TGD Universe is quantum critical. The basic coupling constant strength of quantum TGD is mathematically analogous to critical temperature and has a discrete spectrum. Living matter is expected to be especially interesting quantum critical system.

Under what circumstances is it possible to realize the phase transition transforming ordinary particle to dark matter with non-standard value of Planck constant? This is a question encountered sooner or later by anyone dreaming of building artificial life! Quantum critical fluctuations have long range. On the other hand, quantum coherence length is proportional to $h_{eff} = n \times h$. This suggests the answer: if you want dark matter with $h_{eff} = n \times h$—and maybe even living matter—you must create a quantum critical system! This could be the royal road to the experimentation with dark matter and eventually to a technology controlling it.

The mathematical formulation for this vision involves a generalization of the notion of super-conformal symmetry of super-strings theories. Quantum criticality at the level of super-conformal symmetries in TGD sense implies that the hierarchy of Planck constants $h_{eff} = n \times h$ labels a fractal hierarchy of sub-algebras of super-symplectic and other conformal algebras isomorphic to the full algebra. The physical interpretation is in terms of dark matter hierarchy. One has conformal symmetry breaking without conformal symmetry breaking as Wheeler would have put it. The hierarchy of breakings is possible also for ordinary conformal algebras but for some reason has remained unnoticed.

4.4. Dark matter as a source of long ranged weak and color fields

Long ranged classical electro-weak and color gauge fields are unavoidable in TGD framework. The smallness of the parity breaking effects in hadronic, nuclear, and atomic length scales does not however seem to allow long ranged electro-weak gauge fields. The problem disappears if long range classical electro-weak gauge fields are identified as space-time correlates for massless gauge fields created by dark matter. Also scaled up variants of ordinary electro-weak particle spectra are possible. The identification explains chirality (handedness) selection in living matter (for biomolecules and also for ourselves!) and unbroken $U(2)_{ew}$ invariance and free color in bio length scales become characteristics of living matter and of bio-chemistry and bio-nuclear physics.

The recent view about the solutions of Kähler- Dirac action describing fermions assumes that the modes have a well-defined em charge and this implies that localization of the modes to 2-D surfaces (right-handed neutrino is an exception). Classical W boson fields vanish at these surfaces and also classical Z^0 field can vanish. The latter would guarantee the absence of large parity breaking effects above intermediate boson scale scaling like h_{eff}. There are more general justifications for this assumption from both strong form of holography and number theoretic vision.

5. Magnetic body as a sensory perceiver and intentional agent

The hypothesis that dark magnetic body serves as an intentional agent using biological body as a motor instrument and sensory receptor is consistent with Libet's findings about strange time delays of consciousness.[172] Neuronal activity begins before the conscious decision—understanding this requires ZEO. Also our sensory data is not from recent moment but has age which is fraction of second—as if this information had to travel a length of order Earth size before we receive it.

This suggests that "me" corresponds to my magnetic body with size at least of size of Earth. The double organism-environment would be extended to triple organism-environment-magnetic body. Magnetic body would carry cyclotron Bose-Einstein condensates of various ions. Magnetic body must be able to perform motor control and receive sensory input from biological body. Cyclotron radiation would be one possible communication and control tool.

Cell membrane would be a natural sensor providing information about cell interior and exterior to the magnetic body and dark photons at appropriate frequency range would naturally communicate this information. The strange quantitative coincidences with the physics of cell membrane and high T_c superconductivity support the idea that generalized Josephson radiation generated by generalized Josephson currents of dark electrons through cell membrane is responsible for this communication.[173]

Motor control by magnetic body would be most naturally performed via genome: this is achieved if flux sheets traverse through DNA strands. Flux quantization for large values of Planck constant requires rather large widths for the flux sheets. If flux sheet contains sequences of genomes like the page of book contains lines of text, a coherent gene expression becomes possible at level of organs and even populations and one can speak about super- and hyper-genomes. Introns might relate to the collective gene expression possibly realized electromagnetically rather than only chemically.

Dark cyclotron radiation with photon energy above thermal energy could be used for coordination purposes at least—EEG rhythms. The predicted hierarchy of copies of standard model physics leads to ask whether also dark copies of electro-weak gauge bosons and gluons could be important in living matter.

1. Magnetic bodies would contain flux tubes, which turn back and return to give rise to U-shaped structures— "tentacles". These tentacles would be continually scanning the environment. As the tentacle encounters a tentacle from second system it reconnects with it and under some conditions form a pair of flux tubes connecting the two systems to a single system. Supra currents could flow and resonant cyclotron radiation propagate along the flux tubes. These flux tube pairs would serve as correlates for attention at molecular level. Immune system and even genetic code would have evolved from these tentacles.

2. The shortening of the flux tubes connecting biomolecules in a phase transition reducing Planck constant could be a basic mechanism of bio-catalysis and explain the mysterious ability of biomolecules to find each other.

Similar process in time direction could explain basic aspects of symbolic memories as scaled down representations of actual events.

3. The strange behavior of cell membrane suggests that a dominating portion of important biological ions are actually dark ions at magnetic flux tubes so that ionic pumps and channels are needed only for visible ions. This leads to a model of nerve pulse explaining its unexpected thermodynamical properties with basic properties of Josephson currents making it unnecessary to use pumps to bring ions back after the pulse. The model predicts automatically EEG as Josephson radiation and explains the synchrony of both kHz radiation and of EEG.[174]

4. The DC currents of Becker could be accompanied by Josephson currents running along flux tubes making possible dissipation free energy transfer and quantum control over long distances and meridians of Chinese medicine could correspond to these flux tubes.[175]

5. The model of DNA-cell membrane system[176] as topological quantum computer assumes that nucleotides and lipids are connected by ordinary or "wormhole" magnetic flux tubes acting as strands of braid and carrying dark matter with large Planck constant. The model leads to a new vision about TGD in which the assignment of nucleotides to quarks allows to understand basic regularities of DNA not understood from biochemistry.

6. Each physical system would correspond to an onion-like hierarchy of field bodies characterized by p-adic primes

and value of Planck constant. The highest value of Planck constant in this hierarchy provides kind of intelligence quotient characterizing the evolutionary level of the system since the time scale of planned action and memory correspond to the temporal distance between tips of corresponding causal diamond (CD). Also the spatial size of the system correlates with the Planck constant. This suggests that great evolutionary leaps correspond to the increase of Planck constant for the highest level of hierarchy of personal magnetic bodies. For instance, neurons would have much more evolved magnetic bodies than ordinary cells.

7. At the level of DNA this vision leads to an idea about hierarchy of genomes. Magnetic flux sheets traversing DNA strands provide a natural mechanism for magnetic body to control the behavior of biological body by controlling gene expression. The quantization of magnetic flux states that magnetic flux is proportional to $\hbar$ and thus means that the larger is the value of $\hbar$, the larger is the width of the flux sheet. For larger values of $\hbar$ single genome is not enough to satisfy this condition. This leads to the idea that the genomes of organs, organism, and even population, can organize like lines of text at the magnetic flux sheets and form in this manner a hierarchy of genomes responsible for a coherent gene expression at level of cell, organ, organism and population and perhaps even entire biosphere. This would also provide a mechanism by which collective consciousness would use its biological body—biosphere.

6. Zero Energy Ontology (ZEO) and causal diamonds (CDs)

The evolution of TGD inspired theory of consciousness during the last decade has brought in several new concepts and led to a rather detailed vision making highly non-trivial predictions allowing to test the vision. In particular, the idea that the theory of consciousness follows as a generalization of quantum measurement theory by transforming observer from an outsider to the physical world to a conscious entity described by the laws of (generalized) quantum physics has become rather concrete.

Everything starts from the question "What exists?" Zero energy ontology (ZEO) is a partial answer to this question.

1. ZEO states that physical states have vanishing conserved net quantum numbers and are decomposable to positive and negative energy parts separated by a temporal distance characterizing the system as a space-time sheet of finite size in time direction. The particle physics interpretation is as initial and final states of a particle reaction. This picture is much more general than the ordinary positive energy ontology since any zero energy state can be created from vacuum—this is not in conflict with the field equations and conservation laws. Physics does not exclude free will or vice versa.

2. Zero energy states are superpositions over pairs of positive energy states and negative energy states and correspond to initial and final states of a physical event in the usual positive energy ontology. Positive and negative energy states are localized at the opposite light-like boundaries of a causal diamond (CD) defined essentially as an intersection of future and past directed light-cones

(diamond in 2-D is an excellent visualization of CD). Space-time surfaces in the quantum superposition are identified as preferred extremals of Kähler action and are restricted inside CD for the simplest option.

3. CDs form a fractal hierarchy with size scales coming as integer multiples of a fundamental size scale (CP_2 scale). Translates and Lorentz boosts of CDs are also possible. It is not quite clear whether one should allow CDs to intersect or should one require strict nesting. System has in general wave function in the moduli space of CDs and in quantum jump a localization to CDs for which either upper or lower boundary is fixed takes place.

CDs would serve as the geometric correlates of conscious entities—selves—at the level of imbedding space $M^4 \times CP_2$. The 4-D space-time surfaces define the correlates of selves at space-time level.

7. NMP and the notion of self

The notion of self as conscious entity is central for any theory of consciousness. Quantum measurement theory in ZEO generalizes. What is new is that quantum measurement as state function reduction can occur at both boundaries of CD.

1. Negentropy Maximization Principle (NMP) is the variational principle of consciousness. NMP essentially states that the information content of conscious experience is maximal. Accordingly, the negentropy gain in state function reduction at the active boundary of CD is maximal and also

that the entanglement negentropy at unchanging passive boundary of CD is as large as possible.

NMP is consistent with the standard quantum measurement theory applying when the entanglement is not negentropic: what happens that entanglement is reduced as some subsystem goes to an eigenstate of its density matrix and entanglement entropy is reduced.

2. For negentropic entanglement (NE) the situation changes: NMP does not force the reduction of entanglement now. NMP in fact encourages the generation of NE and NE is rather stable against NMP. This is essential for understanding living systems. There are also obvious applications to quantum computation where the instability of entanglement is the basic problem.

In ZEO self can be seen as a generalized Zeno effect!

1. The idea is simple: self corresponds to a sequence of state function reductions at the same boundary of CD and leaving it and states at it unaffected—call it passive boundary. In ordinary quantum theory this would correspond to Zeno effect, meaning that quantum state remains unaffected (watched kettle does not boil).

2. This is the situation at the passive boundary of CD but at the second—active—boundary states are changing. Active boundary is also gradually shifted to future in the sequence of repeated reductions: this gives rise to the experienced flow of time. The average increase of the temporal distance between the tips of CD defines the life-time of self. The

number of reductions gives a measure for the subjectively experienced life-time of self. Also other sensory experiences come from the change at active boundary of CD and create the world of "Maya". The passive boundary corresponds to the experience about permanent self - sometimes called Self.

3. NMP would eventually force "death" (the reader can decide whether to include the quotation marks) of self since the state function reduction at opposite boundary would generate more negentropy. "Death" of self would mean birth of self asssociated with the opposite boundary of CD. The age of self, identified as the proper time distance between the tips, would increase in statistical sense even when its arrow can change. The act of volition would have a natural identification as the first state function reduction at the opposite boundary of CD.

4. The highly non-trivial prediction is that the arrow of time depends on which boundary of CD is active. As self "dies", a time reversed self is created. This is re-incarnation but not in the usual sense. At the next step something very much analogous to re-incarnation in the ordinary sense occurs. One might speak of transmigration of souls. These re-incarnates have all the negentropy that they managed to gain in earlier lives so that there is genuine evolution involved. This picture also provides a quantal description for the notion of negative energy signals propagating backwards in geometric time serving key role in TGD inspired models of motor action, memory, and remote metabolism—kind of remote eating!

Appendix: Proposal for a P-Adic Mathematics Doctoral Thesis

A primary focus of *Digital Mind Math* is to significantly broaden the potential audience for how Dr. Matti Pitkänen's theory of modern physics—Topological Geometrodynamics, or TGD—offers a promising model for the mind and how we think.

This is challenging generally because TGD is a complex and sophisticated theory of quantum physics and general relativity. And is challenging more specifically because of the role that p-adic mathematics plays.

Although in Digital Mind Math, p-adic mathematics is our first mathematics, our most basic mathematics, it is not taught in elementary school or high school mathematics, and is infrequently taught within the undergraduate mathematics curriculum.

Thus *Digital Mind Math* is breaking into new territory by attempting to make p-adic mathematics intuitively appealing to a general audience.

One risk in attempting this, of course, is that the simplifying becomes oversimplifying, perhaps to the point of being inaccurate.

I think we can all sympathize with the great experts in any field who do not appreciate amateur attemps at oversimplifying their fields. However, it can be argued that any topic, no matter how complicated and advanced, has a topic sentence. And that topic sentence can be broken down into a handful of next-level-down points, and so on. It is not easy to identify and state these simplified essences of p-adic mathematics, but that is what *Digital Mind Math* has attempted to do, both because of a commitment to broaden the audience, and a commitment to the concept that p-adic mathematics is at its heart a simple, intuitive mathematics and the mathematics of how we think, the mathematics of cognition and the mind, our first and most natural mathematics.

Listed below are a set of issues that would benefit from scrutiny and analysis by an expert on advanced concepts of p-adic mathematics. The kind of expert that is needed is one with a commitment to not look to quickly say, "No. That's wrong." Rather, the useful expert would be one with a commitment, where correction or refinement is needed, to do so by preserving the point made as intact as possible, and by offering the most minimal correction possible that changes the point to one that is strictly accurate from the point of view of advanced p-adic analysis, at the same time that the point remains as accessible as possible to a general audience.

This set of mathematically technical issues is Part One in the outline below for a proposed Ph.D. thesis. Part Two of the thesis is the practical implementation of the Digital Mind Math model.

APPLIED P-ADIC MATHEMATICS PH.D. THESIS: IMPLEMENTATION OF DIGITAL MIND MATH

PART ONE: MATHEMATICAL ASSUMPTIONS AND QUESTIONS IMPLICIT WITHIN THE TEXT OF *DIGITAL MIND MATH*

I. **HENSEL'S LEMMA**
    A. Although the original core purpose of Hensel's Lemma was not about creating a p-adic analog to Newton's Method, any version of Hensel's Lemma defines the conditions under which approximations converge to a p-adic root as a by-product of the proof. (Correct this introductory comment if it's not 100% accurate.)
    B. For Digital Mind Math, the main application asserted for Hensel's Lemma relates to maximization of the p-adic norm (rather than to finding the root of an equation). Is this a valid application of Hensel's Lemma?

1. In other words, the application of Hensel's Lemma within Digital Mind Math is to take the conditions that Hensel's Lemma specifies as conditions for the consequence that: If an action or step increases the p-adic norm locally, it's guaranteed to increase the p-adic norm globally. (This is as opposed to taking the conditions that Hensel's Lemma specifies as conditions for the consequence that we're converging to a root.)

2. Is there any sense in which a weaker condition than always increasing the p-adic norm locally will still guarantee the increase of the p-adic norm globally? Specifically, can a set of globally increasing local structural conditions be specified to guarantee that the p-adic norm is globally increased, even if the immediate local action temporarily decreases the p-adic norm? (This would be similar to the concept in finance of the present value of future cash flows, a quantity that could increase because cash flows increase in the long term, even if they are initially depressed by capital investments.)

C. The simplest version of Hensel's Lemma (with the conditions that $f(a_0) \equiv 0$ (mod p), and $f'(a_0)$ *not*$\equiv 0$ (mod p)), is a special case of the more general conditions that $f(a_0) \equiv 0$ (mod p^{2M+1}), and $f'(a_0)$ *not*$\equiv 0$ (mod p^{M+1}), and $f'(a_0) \equiv 0$ (mod p^M). Other versions of Hensel's Lemma relate to the p-adic norm quantity $|a - a_0|_p$. How do statements of Hensel's Lemma in terms of the p-adic norm relate to statement in terms of $f(a_0)$ and $f'(a_0)$?

D. Part Four of *Digital Mind Math* centers on mathematician Keith Conrad's "tightened" version of Hensel's Lemma, which "provides a converse of sorts" to Hensel's Lemma.[177]

1. Conrad's paper is stated in terms of the non-Archimedean norm in general, rather than specifically the p-adic norm. In Digital Mind Math, due to Ostrowski, we are interested in only the p-adic norm specifically. Is it accurate to apply Conrad's work specifically to the p-adic norm throughout?

2. Are there other statements of a converse to Hensel's Lemma—that is, other statements of the tightened conditions which are the necessary (not just sufficient) conditions under which Hensel's Lemma holds?

3. How would Conrad's analysis change without the restrictions that $|f(a_0)| \leq 1$ and that $|f'(a_0)| \leq 1$?

II.   WITT VECTORS

A. What are the elements of p-adic Witt vectors: P-adic numbers? (Fully realized as the sum of Teichmüller coefficients times rational powers of p?) Or p-adic digits (Teichmüller elements)?

B. The main question regarding Witt vectors for Digital Mind Math is how Witt vectors offer enhanced power for manipulating large quantities of data. Specifically, we are looking to mathematically manipulate not just sequences of p-adic digits, but rather sequences of p-adic numbers. (The only operation that we're interested in performing on these sequences of p-adic numbers is maximization—not, for example, addition or multiplication—so you may take advantage of this if it makes the answer any easier.)

1. If the elements of p-adic Witt vectors are p-adic numbers (rather than just p-adic digits), then this in

and of itself offers enhanced power, compared to p-adic numbers alone.

2. Can it be said that the set (ring? field?) of Witt vectors is homomorphic (isomorphic?) to the set (ring? field?) of p-adic numbers?

3. How is it (explained simply, intuitively) that the Frobenius endomorphism, the Verschiebung operation, and the restriction map operate to offer this enhanced power?

4. Are there other aspects of Witt vectors that contribute to this enhanced power?

C. The main quantification of p-adic numbers within Digital Mind Math is the p-adic norm (p-adic absolute value). How is this calculated for universal Witt vectors?

D. In the fully complete and closed p-adic space Ω ,one formulation of p-adic numbers is as the sum of Teichmüller coefficients times rational powers of p. (Scott Carnahan attributes this formulation to Bjorn Poonen's undergraduate thesis.[178])

1. Can Witt vectors be formulated to operate in this fully complete and closed p-adic space Ω ?

2. There is the same infinite number $\aleph_0$ of integers as there is of rational numbers. So why is there any enhanced scope if we can raise p to rational powers than there is if we're constrained to raising p to integer powers?

E. Does a de Rahm-Witt complex construction of sheaves of Witt vectors achieve the enhanced power of p-adic mathematics that we're seeking?

1. Of course, it is the big de Rham-Witt complex, not just the p-typical de Rham-Witt complex, that is of interest.

PART TWO: RIGOROUSLY DESIGN THE FULL DIGITAL MIND MATH PROCESS

I. **THE ORGANIZATION OF THE MIND**

    A. Construct a sample space of intricately linked concepts, which can be traveled along various paths to form various thoughts.

        1. This is Topological Geometrodynamics' and Digital Mind Math's $M^4_+ \times CP_2$ space, so core M^4_+ video segments are linked to many descriptive tags, with intricate partial overlap among the various videos' taggings.

    B. Create the p-adic labeling for the $M^4_+ \times CP_2$ structure.

II. **THE CORE COGNITIVE PROCESS**

    A. Select as a starting point a sequence from the $M^4_+ \times CP_2$ structure that represents the prior thought.

    B. Using the p-adic representation of this starting point, generate the p-adic representation of all possible next thoughts.

    C. Select the next thought to experience based on the maximization of the p-adic norm.

    D. Identify how the p-adic label of the selected thought relates to the real mathematics of the thought as it is experienced. (For this purpose, Topological Geometrodynamics suggests that the p-adic/real correspondence is achieved based on common rationals, or, more expansively stated, intersecting at common rationals of TGD's world of classical worlds, rather than simply its points.)

III. THE IDEAL MIND

A. Apply the converse of Hensel's Lemma to Digital Mind Math, in order to draw conclusions about how thought is not fully free-form, but rather: the selection of possible next thoughts can be circumscribed within a specific limited set of choices.

1. This will require developing an intuitive formulation of how the derivative of a p-adic function relates to the Digital Mind Math model.

2. In addition, Conrad's equivalent formulation, in terms of the p-adic norm $|a - a_0|_p$, is of interest: How can this formulation be intuitively applied within the Digital Mind Math framework, perhaps especially to define a Zone of Proximal Development?

B. A number of theoretical issues need to be resolved to extend this beyond Conrad's statement of the converse of Hensel's Lemma:

1. Can this be extended beyond a p-typical framework to a framework of universal Witt vectors?

   a. This would introduce a competing way to increase the p-adic norm, besides (in a framework of p-adic integers) increasing the p-adic norm by decreasing n. This second way to increase the p-adic norm is (in a framework of p-adic integers) by decreasing p, for which universal Witt vectors open the door.

2. And further extended, to the big de Rham-Witt complex framework?

Selected Bibliography

Listed here are technical references that underlie the mathematics and science discussed throughout *Digital Mind Math*. In addition, full bibliographic information appears in the endnotes for specific quotes and other source citations.

Topological Geometrodynamics
Pitkänen, Matti. *Topological Geometrodynamics.* www.tgdtheory.fi.

------. *Topological Geometrodynamics.* Luniver Press, 2006.

------. *Life and Consciousness: TGD Based Vision*. LAP LAMBERT
        Academic Publishing, 2014.

------. *TGD diary.* www.matpitka.blogspot.com.

Paster, Robert. *New Physics and the Mind*. BookSurge, 2006.

P-Adic Mathematics.
Carnahan, Scott. *p-adic fields for beginners.*
        http://sbseminar.wordpress.com/2007/08/21/p-adic-fields-
        for-beginners.

Hazewinkel, Michael. "Witt vectors. Part 1."
        http://arxiv.org/ftp/arxiv/papers/0804/0804.3888.pdf. April
        20, 2008.

Katok, Svetlana. *p-adic Analysis Compared with Real.* American
        Mathematical Society, 2007.

Koblitz, Neal. *p-adic Numbers, p-adic Analysis, and Zeta-Functions.*
        Springer, 1977.

Cognitive Development
Ginsburg, Herbert, and Sylvia Opper. *Piaget's Theory of Intellectual
        Development: An Introduction*. Prentice-Hall, 1969.

Inhelder, Bärbel, and Jean Piaget. *The Growth in Logical Thinking from
        Childhood to Adolescence*. Basic Books, 1958.

Piaget, Jean, and Bärbel Inhelder. *The Psychology of the Child*. Basic
     Books, 1969.

Vygotsky, L.S. "Play and Its Role in the Mental Development of the
     Child." *Voprosy psikhologii* 12 (6), 1966, 62-76. Retrieved from
     http://www2.winchester.ac.uk/edStudies/arch12-
     13/level%20two%20sem%20one/es2212w11%20v2.pdf,
     uncredited translator and editor.

Vygotsky, L.S., Michael Cole, Vera John-Steiner, Sylvia Scribner, and
     Ellen Souberman. *Mind in Society: The Development of Higher
     Psychology Processes*. Harvard University Press, 1980.

Vygotsky, Lev S. and Alex Kozulin. Thought and Language. M.I.T. Press,
     2012.

Endnotes

[1] American Dialect Society, "2015 Word of the Year is singular 'they.'" http://www.americandialect.org/2015-word-of-the-year-is-singular-they, 8 January 2016.

[2] In Molière's 17th-century play *The Bourgeois Gentleman*, the protagonist Monsieur Jourdain expresses pleasure when he finds out he's been speaking prose all his life without knowing it.

[3] The discussion of Piaget draws from: Herbert Ginsburg and Sylvia Opper, *Piaget's Theory of Intellectual Development: An Introduction*, Prentice-Hall, 1969; Bärbel Inhelder and Jean Piaget, *The Growth in Logical Thinking from Childhood to Adolescence*, Basic Books, 1958; and Jean Piaget and Bärbel Inhelder, *The Psychology of the Child*, Basic Books, 1969.

[4] The discussion of Vygotsky draws from: L. S. Vygotsky, "Play and Its Role in the Mental Development of the Child," *Voprosy psikhologii* 12 (6), 1966, 62-76, retrieved from http://www2.winchester.ac.uk/edStudies/arch12-13/level%20two%20sem%20one/es2212w11%20v2.pdf, uncredited translator and editor; L. S. Vygotsky, Michael Cole, Vera John-Steiner, Sylvia Scribner, and Ellen Souberman, *Mind in Society: The Development of Higher Psychology Processes*, Harvard University Press, 1980; and Lev S. Vygotsky and Alex Kozulin, *Thought and Language*, M.I.T. Press, 2012.

[5] Sam Byford, "Google's AlphaGo AI beats Lee Se-dol again to win Go series 4-1." *The Verge*, http://www.theverge.com/2016/3/15/11213518/alphago-deepmind-go-match-5-result, 15 March 2016.

[6] Susana F. Huelga and Martin B. Plenio, "Quantum biology: A vibrant environment." *Nature Physics*, vol. 10 (2014), pp. 621-622.

[7] M. Ebara et al., "Smart Biomaterials." Springer NIMS Monographs, www.springer.com/materials/biomaterials/book/978-4-431-54399-2 (accessed 24 June 2014).

[8] Tom Simonite, "Carbon Nanotubes Could Step In to Sustain Moore's law." *MIT Technology Review*, vol. 117, no. :5 (2014), p. 17.

[9] Tom Simonite, "Microsoft's Quantum Search for the 'Next Transistor.'" *MIT Technology Review*, vol. 117, no. :5 (2014), p. 21.

[10] Yi-Kai Liu, "Quantum information: Show, don't tell." *Nature Physics*, vol. 10 (2014), pp. 625-626.

[11] Eric Smalley, "Quantum Engineering." *Spectrum*, Spring 2014, p. 17.

[12] Courtney Humphries, "Brain Mapping." *MIT Technology Review*, www.technology review.com/featuredstory/526501/brain-mapping (accesed 1 September 2014).

[13] Jonathan Webb, "Brain-inspired chip fits 1m 'neurons' on postage stamp." BBC News, www.bbc.com/news/science-environment-28688781 (8 August 2014).

[14] Fergus Walsh, "Billion pound brain project under way." BBC News, www.bbc.com/news/health-24428162 (7 October 2013).

[15] In Scott Carnahan, "p-adic fields for beginners," http://sbseminar.wordpress.com/2007/08/21/p-adic-fields-for-beginners (21 August 2007), Carnahan attributes this formulation to Bjorn Poonen.

[16] Scott Carnahan, "p-adic fields for beginners." http://sbseminar.wordpress.com/2007/08/21/p-adic-fields-for-beginners (21 August 2007).

[17] Michael Hazewinkel, "Witt vectors. Part 1," p. 7. http://arxiv.org/abs/0804.3888. (20 April 2008).

[18] Ibid.

[19] Ibid, p. 127.

[20] Luc Illusie, "Crystalline Cohomology." *Proceedings of Symposia in Pure Mathematics*, vol. 55, part 1 (1994), pp. 43-70.

[21] Paul VanKoughnett, "Crystalline cohomology and de Rham-Witt complex." http://math.northwestern.edu/~dwilson/k3notes/Lecture21-Crystalline.pdf (12 November 2014).

22 "Idea of the de Rham-Witt complex."
http://www3.nd.edu/~mbehren1/TAGS/Davis_notes.pdf
(author not identified).

23 Lars Hesselholt, "The big de Rham-Witt complex." *Acta
Mathematica*, vol. 214, iss. 1 (March 2015), pp. 135-207.

24 Jean Piaget, "The Right to Education in the Modern World."
Freedom and Culture, UNESCO (Columbia University Press,
1951), pp. 67-116.

25 Susan Conova, "New Clues About Synapse Formation." *In Vivo* vol.
2, iss. 14 (15 September 2003), Columbia University Health
Sciences: http://www.cumc.columbia.edu/publications/in-
vivo/Vol2_Iss14_sept15_03/research-briefs.html

26 "About The Music Genome Project." www.pandora.com/about/mgp

27 Will Knight, "The Hit Charade." *MIT Technology Review*, vol. 118, no.
6 (November/December 2015), pp. 79-82.

28 Ibid.

29 Jenny Hendrix, "Illusion," review of *The Confabulist*, by Steven
Gallowa. *The New York Times Book Review*, 1 June 2014, p. 45.

30 Adam Bryant, "Ron Kaplan of Trex, on Making Judgments Instead of
Decisions." Corner Office column, *New York Times*, 3 May
2014, SundayBusiness p.2.

31 Ofer Zur, "To Cross Or Not To Cross: Do Boundaries In Therapy
Protect Or Harm." *Psychotherapy Bulletin*, vol. 39, iss. 3
(2004), pp. 27-32.

32 "10 Ways to Build and Preserve Better Boundaries."
psychcentral.com/lib/10-way-to-build-and-preserve-better-
boundaries/ (accessed 22 January 2016).

33 "Setting Boundaries with Difficult People."
ipfw.edu/affiliates/assistance/selfhelp/relationship-
settingboundaries.html (acessed 22 January 2016).

34 "6 Steps to Setting Boundaries in Relationships."
jennifertwardowski.com/2014/06/23/boundaries-in-
relationships (accessed 22 January 2016).

35 "How to Set Healthy Boundaries: 3 Crucial First Steps."
tinybuddha.com/blog/how-to-set-healthy-boundaries-3-
crucial-first-steps (accessed 22 January 2016).

[36] "4 Ways to Establish Boundaries." wikihow.com/Establish-Boundaries (accessed 22 January 2016).

[37] Raymond Lloyd Richmond, "A Guide to Psychology and Its Practice: Boundaries." guidetopsychology.com/boundaries.htm

[38] Laverne Cox as quoted in Erik Piepenburg, "Longing for the Door to Swing Fully Open." *The New York Times*,21 June 2015, Arts & Leisure pp. 12-13.

[39] Mary Gordon, "Many Splendored Things," review of Rose Tremain's *The American Lover and Other Stories*. *The New York Times Book Review*, 1 March 2015, p. 10.

[40] Rachel Shteir, "Publish and Cherish," review of Kevin Birmingham's *The Most Dangerous Book: The Battle for James Joyce's "Ulysses."* *The New York Times Book Review*, 24 August 2014, p. 27.

[41] Stephen Yablo, *Aboutness*. Princeton University Press, 2014.

[42] The discussion of Piaget draws from: Herbert Ginsburg and Sylvia Opper, *Piaget's Theory of Intellectual Development: An Introduction*, Prentice-Hall, 1969; Bärbel Inhelder and Jean Piaget, *The Growth in Logical Thinking from Childhood to Adolescence*, Basic Books, 1958; and Jean Piaget and Bärbel Inhelder, *The Psychology of the Child*, Basic Books, 1969.

[43] John Gray, *Men Are from Mars, Women Are from Venus*. Harper, 1993.

[44] Christopher Jones, "Examples of Piagetian Assimilation and Accommodation." webspace.pugetsound.edu/cjones/piaget-examples.doc

[45] Chimamanda Ngozi Adichie, "The danger of a single story." TEDGlobal 2009, http://www.ted.com/talks/chimamanda_adichie_the_danger_of_a_single_story?language=en (July 2009).

[46] Ibid.

[47] Sally Kohn, "Waco coverage shows double standard on race." www.cnn.com/2015/05/18/opinions/kohn-biker-shooting-waco/, 19 May 2015.

[48] Alex Pentand, *Social Physics: How Good Ideas Spread—The Lessons from a New Science*. Penguin Press, 2014.

[49] Adam Bryant, "Making Room for Differences." *The New York Times*, 8 February 2015, SundayBusiness p. 2.

[50] Colson Whitehead, "Life Without Pity." *The New York Times Magazine*, 8 March 2015 pp. 17-19.

[51] Valve Games, www.valvesoftware.com/games/portal.html

[52] Ryan Faith, "Figuring Out the Future of War in the Pacific—Or, What the Hell is Seabasing." www.news.vice.com/article/figuring-out-the-future-of-war-in-the-pacific-or-what-the-hell-is-seabasing

[53] Julio E. Correa, "The Dog's Sense of Smell." Alabama Cooperative Extension System, www.aces.edu/pubs/docs/U/UNP-0066/UNP-0066.pdf

[54] David Alexander Smith, *In the Cube*. Tor Books, 1994.

[55] Peter Dizikes, "It's About Time," reviewing Bradford Skow, *Objective Becoming*, Oxford University Press, 2015. *MIT Technology Review*, March/April 2015, p. MIT News 10.

[56] Ibid.

[57] Margarita Vázquez and Antonio Manuel Liz Gutiérrez (eds.), *Temporal Points of View: Subjective and Objective Aspects*. Springer Studies in Applied Philosophy, Epistemology, and Rational Ethics, 2015.

[58] Sam Anderson, "Standstill." *The New York Times Magazine*, 30 August 2015, pp. 15-17.

[59] Paul Krugman, "The Powers That Were." *The New York Times Book Review, 31* January 2016, pp. 1, 18.

[60] Peggy Orenstein, "The Wrong Approach to Breast Cancer." *NewYork Times*, 27 July 2014, Sunday Review p. 4.

[61] David Robson, "Neuroscience: The man who saw time freeze." *BBC Future: In Depth*, bbc.com/future/columns/in-depth (June 25, 2014).

[62] Robert S. Pingree, "Mrs. Earp." *The New York Times Book Review, 13 July 2014, p. 5.*

[63] Ibid.

[64] *Love and Death*. directed by Woody Allen. United Artists, 1975. Film.

[65] *Dressed to Kill*, directed by Roy William Neill, 1946 (Vintage Home Entertainment DVD, 2003).

[66] "P.D. James, Talking and Writing 'Detective Fiction.'" Interview with Linda Wertheimer, NPR (22 December 2009).

[67] Tim Dirks, "Greatest Film Plot Twists, Film Spoilers and surprise Endings." AMC filmsite: www.filmsite.org/greattwists.html.

[68] Ibid.

[69] #10 at Donald Deane, ""The 10 Most Confusing Movies of All Time." *Screen Crush*, http://screencrush.com/confusing-movies (30 May 2013).

[70] #4 at "The Most Confusing Movies Ever Made." *Ranker*, http://www.ranker.com/list/most-confusing-movies-ever-made/ranker-film (accessed 3 February 2016).

[71] #1 at tanikos 1311, "The 99 Confusing Films Ever Made." IMDb, http://www.imdb.com/list/ls000135739 (15 April 2011).

[72] #4 at Daniel Flynn, "10 Most Confusing Films of Al-Time." *WhatCulture*, http://whatculture.com/film/10-most-confusing-films-of-all-time.php (accessed 3 February 2016).

[73] #6 at "15 of the Most Confusing Films Ever Made." *PopCrunch*, http://www.popcrunch.com/most-confusing-films/?img=131543 (accessed 3 February 2016).

[74] David Lynch, "David Lynch's 10 Clues to Unlocking the Thriller." Card included with *Mulholland Drive* (DVD). Universal Focus. 2004.

[75] Kate Murphy, "Download" column, interviewing Peggy Smith. *New York Times*, 15 June 2014, *Sunday Review* p. 2. Smith was discussing her enjoyment of watching Masterpiece mysteries, such as Poirot, Dr. Finlay, Inspector Morse, and Jane Marple.

[76] César Aira, *An Improvisational Mind*. New Directions, 2015. As quoted in Patti Smith, "An Informational Mind," *The New York Times Book Review*, 15 March 2015, p. 10.

[77] Karen Armstrong, "Articles of Faith." *The New York Times Book Review*, 21 December 2014, p. 9, quoting Jack Miles in her review of Jack Miles, David Biale, Lawrence S. Cunningham, and Jane Dammen McAuliffe (eds.), *The Norton Anthology of*

World Religions, Volume II: Judaism, Christianity, Islam, W.W. Norton & Company, 2014.

[78] Susan Engel, *The Hungry Mind: The Origins of Curiosity in Childhood*. Harvard University Press, 2015.

[79] Nancy Andreasen, "Secrets of the Creative Brain." *The Atlantic*, 25 June 2014, theatlantic.com/features/archive/2014/06/secrets-of-the-creative-brain/372299/

[80] Ibid.

[81] Kelly Clancy, "YourBrain Is On the Brink of Chaos." *Nautilus*, Issue 15: Turbulence, nautil.us/issue/15/turbulence/your-brain-is-on-the-brink-of-chaos

[82] Audie Cornish, "For a Year, Shonda Rhimes Said 'Yes' To All The Things That Scared Her," interview with Shonda Rhimes. NPR,9 November 2015, npr.org/2015/11/09/455340952/for-one-year-shonda-rhimes-said-yes-to-all-the-things-that-scared-her, discussing Shonda Rhimes' memoir *Year of Yes: How to Dance It Out, Stand in the Sun and Be Your Own Person*, Simon & Schuster, 2015.

[83] Christopher Lasch, *The Culture of Narcissism: American Life in an Age of Diminishing Expectations*. W.W. Norton, 1979.

[84] Rebecca Saxe, as interviewed in Courtney Humphries, "What Am I Thinking About You." *MIT Technology Review*, vol. 117, no. 4 (July/August 2014), pp. 60-61.

[85] Michael S. Graziano, "Are We Realy Conscious?" *New York Times*, 12 October 2014, Sunday Review p. 12.

[86] Ibid.

[87] The idea that it is information that is the most fundamental entity of the universe is discussed, for example, by Manoj K. Samal in "Speculations on a Unified Theory of Matter and Mind," xxx.lanl.gov/physics/0111035 (8 November 2001) and "Can Science 'Explain' Consciousness," xxx/lan1.gov/physics/0002045 (24 February 2000).

[88] Neal Koblitz, "p-*adic Numbers, p-adic Analysis, and Zeta-Functions* (Second Edition), p. 18. Springer, 1984.

[89] Svetlana Katok, p-*adic Analysis Compared with Real*, p. 36. American Mathematical Society, 2007.

[90] Keith Conrad, "Tightening the Basic Version of Hensel's Lemma." http://citeseerx.ist.psu.edu/viewdoc/download?doi=10.1.1.21 0.7490&rep=rep1&type=pdf.

[91] Ana Marie Cox, "Bob Odenkirk Thinks Comedians Are Lousy Critics." *New York Times Magazine*, 24 January 2016, p. 68, quoting Bob Odenkirk.

[92] *Star Trek Into Darkness*. directed by J. J. Abrams. Paramount Pictures, 2013. Film.

[93] Rolling Stones, "You Can't Always Get What You Want." MetroLyrics, http://www.metrolyrics.com/you-cant-always-get-what-you-want-lyrics-rolling-stones.html, accessed 17 March 2016.

[94] Hall & Oates, "I Can't Go For That (No Can Do)." www.azlyrics.com/lyrics/halloates/icantgoforthatnocando.ht ml, accessed 11 April 2016.

[95] *The Chicago Manual of Style* (15th ed.). The University of Chicago Press, 2003.

[96] Ibid., p. 955.

[97] Ibid., p. 213.

[98] Ibid., pp. 755-801.

[99] Ibid., p. 766.

[100] Ibid., pp. 766-767.

[101] Ibid., p. 756.

[102] "'I Regret Everything': Toni Morrison Looks Back On Her Personal Life," Terri Gross interview with Toni Morrison. *NPR Fresh Air,* 20 April 2015, http://www.npr.org/2015/04/20/400394947/i-regret-everything-toni-morrison-looks-back-on-her-personal-life

[103] Ibid.

[104] Alcoholics Anonymous, "The Twelve Steps Of Alcoholics Anonymous." http://www.aa.org/assets/en_US/smf-121_en.pdf

[105] Roy Hoffman, "The Country of Old Age," *The New York Times Book Review*, 14 February 2016, p. 11, quoting from Diane Athill's

Alive, Alive Oh!: And Other Things That Matter (W.W. Norton & Company, 2016).

[106] Stefano Salvatore, "Optical Metamaterials by Block Copolymer Self-Assemby." Springer eBooks: Springer Theses, 2015.

[107] Kohtaro Osakada (ed.), "Organometallic Reactions and Polymerization." Springer eBooks: Lecture Notes in Chemistry, vol. 85, 2014.

[108] P. Gowtham Srinivas, P. Krishna Reddy, A.V. Trinath, S. Bhargav, and R. Uday Kiran, "Mining coverage patterns from transactional databases." *Journal of Intelligent Information Systems*, vol. 45, iss. 3 (December 2015), pp. 423-439.

[109] Marwan Hassani, Yunsu Kim, Seungjin Choi, and Thomas Seidl, "Subspace clustering of data streams: new algorithms and effective evaluation measures." *Journal of Intelligent Information Systems*, vol. 45, iss. 3 (December 2015), pp. 319-335.

[110] Guoyin Wang and Ji Xu, "Granular computing with multiple granular layers for brain big data processing." *Brain Informatics*, vol. 1, iss. 1 (December 2014), pp. 1-10.

[111] Shifei Ding, Mingjing Du, and Hong Zhu, "Survey on granularity clustering." *Cognitive Neurodynamics*, vol. 9, iss. 6 (December 2015), pp. 561-572.

[112] Guandong Xu, Yu Zong, Ping Jin, Rong Pan, and Zongda Wu, "KIPTC: a kernel information propagation tag clustering algorithm." *Journal of Intelligent Information Systems*, vol. 45, iss. 1 (August 2015), pp. 95-112.

[113] Albrecht Zimmermann, "Objectively evaluating condensed representations and interestingness measures for frequent itemset mining." *Journal of Intelligent Information Systems*, vol. 45, iss. 3 (December 2015), pp. 299-317.

[114] D. Fox Harrell, *Phantasmal Media: An Approach to Imagination, Computation, and Expression*. MIT Press, 2013.

[115] Inés Couso, Didier Dubois, and Luciano Sánchez, *Random Sets and Random Fuzzy Sets as Ill-Perceived Random Variables*. Springer eBook, 2014.

[116] Jing Jane He, Daniel Panario, Qiang Wang, and Arne Winterhof, "Linear complexity profile and correlation measure of interleaved sequences." *Cryptography and Communications*, vol. 7, iss. 4 (December 2015), pp. 497-508.

[117] G. Iurato and A. Khrennikov, "Hysteresis model of unconscious-conscious interconnection: Exploring dynamics on *m*-adic trees." *P-Adic Numbers, Ultrametric Analysis, and Applications*," vol. 7, iss. 4 (October 2015), pp. 312-321.

[118] Guilherme Perin, Laurent Imbert, Philippe Maurine, and Lionel Torres, "Vertical and horizontal correlation attacks on RNS-based exponentiations." *Journal of Cryptographic Engineering*, vol. 5, iss. 3 (September 2015), pp. 171-185.

[119] Allen Q. Ye et al., "The intrinsic geometry of the human brain connectome." *Brain Informatics*, vol. 2, iss. 4 (December 2015), pp. 197-210.

[120] Jennifer E. Padilla, Wenyan Liu, Nadrian C. Seeman, "Hierarchical self assembly of patterns from the Robinson tilings: DNA tile design in an enhanced Tile Assembly Model." *Natural Computing*, vol. 11, iss. 2 (June 2012), pp. 323-338.

[121] S. V. Kozyrev, "Ultrametricity in the theory of complex systems." *Theoretical and Mathematical Physics*, vol. 185, iss. 2 (November 2015), pp. 1665-1677.

[122] Haiguang Wen and Zhongming Liu, "Separating Fractal and Oscillatory Components in the Power Spectrum of Neurophysiological Signal." *Brain Topography*, vol. 29, iss, 1 (January 2016), pp. 13-26.

[123] Soumya Sen, Agostino Cortesi, and Nabendu Chaki, *Hyper-lattice Algebraic Model for Data Warehousing*. SpringerBriefs in Applied Sciences and Technology, 2016.

[124] Frank Swain, "Robotics: How machines see the world." *BBC Future In Depth*, www.bbc.com/future/columns/in-depth (25 August 2014). Swain is quoting Peter McOwan of Queen Mary University of London.

[125] Ibid.

[126] Christopher Packham, "New model describes cognitive decision making as the collapse of a quantum superstate."

http://m.phys.org/news/2015-08-cognitive-decision-collapse-quantum-superstate.html (12 August 2015).

[127] Diederik Aerts, Sandro Sozzo, and Tomas Veloz, "Quantum Structure in Cognition and the Foundations of Human Reasoning." *International Journal of Theoretical Physics*, vol. 54, iss. 12 (December 2015), pp. 4557-4569

[128] Tomi H. Johnson, Stephen R. Clark, and Dieter Jaksch, "What is a quantum simulator?" *EPJ Quantum Technology*, vol. 1 (2014), p. 10.

[129] Lee A. Rozema, Dylan H. Mahler, Alex Hayat, Peter S. Turner, and Aephraim M. Steinberg." Quantum Data Compression of a Qubit Ensemble," *Physical Review Letters*, vol. 113, p. 160504 (17 October 2014).

[130] Jack Hansom *et al.*, "Environment-assisted quantum control of a solid-state spin via coherent dark states." *Nature Physics*, vol. 10 (2014), pp. 725-730.

[131] F. Gertz, A. Kozhevnikov, Y. Filimonov, and A. Khitun, "Magnonic Holographic Memory." arXiv:1401.5133 [cond-mat] (21 January 2014).

[132] Adrien Signoles *et al.*, "Confined quantum Zeno dynamics of a watched atomic arrow." *Nature Physics*, vol. 10 (2014), pp. 715-719.

[133] Jonathan Webb, "'Twisted light' beamed across Vienna." *BBC News Science & Environment*, www.bbc.com/news/science-environment-29953239 (11 December 2014).

[134] Eolo DiCasola, "Between Quantum and Classical Gravity: Is There a Mesoscopic Spacetime." *Foundations of Physics*, vol. 45, iss. 2 (December 20, 2014), pp. 171-176.

[135] Diederick Aerts, Sandro Sozzo, and Tomas Veloz, "The Quantum Nature of Identity in Human Thought: Bose-Einstein Statistics for Conceptual Indistinguishability." *International Journal of Theoretical Physics*, vol. 54, iss. 12 (December 2015), pp. 4430-4443.

[136] M. L. Dalla Chiara, R. Giuntini, and E. Negri, "A Quantum Approach to Vagueness and to the Semantics of Music." *International*

     Journal of Theoretical Physics, vol. 54, iss. 12 (December 2015), pp. 4546-4556.

[137] Jaroslaw Pykacz, "Can Many-Valued Logic Help to Comprehend Quantum Processes." *International Journal of Theoretical Physics*, vol. 54, iss. 12 (December 2015), pp. 4367-4375.

[138] Ulrik L. Andersen, Jonas S. Neergaard-Nielsen, Peter van Loock, and Akira Furusawa, "Hybrid discrete- and continuous-variable quantum information." *Nature Physics*, vol. 11 (1 September 2015), pp. 713-719.

[139] Giacomo M. D'Ariano and Paolo Perinotti, "Quantum Theory is an Information Theory." *Foundations of Physics*, vol. 46, iss. 3 (March 2016), pp. 269-281.

[140] "Quantum Computing and Devices (QCD)'" Aalto University School of Science, Department of Applied Physics, Mikko Möttönen (Group Leader), http://physics.aalto.fi/en/groups/qcd/

[141] Masanari Asano, Irina Basieva, Andrei Khrennikov, Masanori Ohya, Yoshiharu Tanaka, and Ichiro Yamato, "Quantum Information Biology: From Information Interpretation of Quantum Mechanics to Applications in Molecular Biology and Cognitive Psychology." *Foundations of Physics*, vol. 45, iss. 10 (October 2015), pp. 1362-1378.

[142] Wei Chen, Guangchuan Wang, and Ruikang Tang, "Nanomodification of living organisms by biomimetic mineralization." *Nano Research*, vol. 7, iss. 10 (October 2014), pp. 1404-1428.

[143] Anirban Samanta, Zhengtao Deng, Yan Liu, and Hao Yan, "A perspective on functionalizing colloidal quantum dots with DNA." *Nano Research* vol. 6, iss. 12 (December 2013), pp. 853-870.

[144] Research of Mansoo Choi and other South Korean engineers, as reported by Jonathan Webb in "Spider-style sensor detects vibrations." *BBC News Science & Environment*, www.bbc.com/news/science-environment-30414752 (11 December 2014).

[145] K. V. Shaitan and I. A. Orshanskiy, "The molecular dynamics of the self-assembly and a rheological model of the superhelical structure of a spiderweb protofibril." *Molecular Biophysics*, vol. 60, iss. 4 (July 2015), pp. 538-541.

[146] Paul Zelisko (ed.), *Bio-Inspired Silicon-Based Materials*. Springer: 2014.

[147] Work of MIT Media Lab and McGovern Institute for Brain Research scientist Ed McGovern, as reported by Cathryn Delude, " Controlling the Brain with Light." *MIT Spectrum*, Fall 2014.

[148] I.P.Pavlova, "Central Pattern Generators." *Neuroscience and Behavioral Physiology*, vol. 45, iss. 1 (January 2015), pp. 42-57.

[149] B. D. Evans, "STDP in lateral connections creates category-based perceptual cycles for invariance learning with multiple stimuli." *Biological Cybernetics*, vol. 109, iss. 2 (April 2015), pp. 215-239.

[150] Patricia Melin, Oscar Castillo, and Janusz Kacprzyk (eds.), *Design of Intelligent Systems Based on Fuzzy Logic, Neural Networks and Nature-Inspired Optimization*. Springer, 2015.

[151] P. A. Merolla, J. V. Arthur, *et al.*, "A million spiking-neuron integrated circuit with a scalable communication network and interface." *Science*, vol. 345, iss. 6197 (8 August 2014), pp. 668-673.

[152] Frits Dannenberg, Marta Kwiatkowska, Chris Thachuk, and Andrew J. Turberfield, "DNA walker circuits: computational potential, design, and verification." *Natural Computing*, vol. 14, iss. 2 (June 2015), pp. 195-211.

[153] Rizki Mardian and Kosuke Sekiyama, "Ant Systems-Based DNA Circuits." *BioNanoScience*, vol. 5, iss. 4 (December 2015), pp. 206-216.

[154] Nazmul Siddique and Hojjat Adeli, "Nature Inspired Computing: An Overview and Some Future Directions." *Cognitive Computation*, vol. 7, iss. 6 (December 2015), pp. 706-714.

[155] http://tgdtheory.fi/appfigures/manysheeted.jpg

[156] http://tgdtheory.fi/appfigures/wormholecontact.jpg

[157] https://en.wikipedia.org/wiki/ER=EPR

[158] https://en.wikipedia.org/wiki/P-adic_number

[159] https://en.wikipedia.org/wiki/Adele

[160] http://tgdtheory.fi/public_html/padphys/padphys.html

[161] http://tgdtheory.fi/public_html/articles/harmonytheory.pdf

[162] https://en.wikipedia.org/wiki/Holographic_principle

[163] https://en.wikipedia.org/wiki/AdS/CFT_correspondence

[164]

http://tgdtheory.fi/public_html/neuplanck/neuplanck.html#qcritdark

[165] http://arxiv.org/abs/astro-ph/0310036

[166] http://tgdtheory.fi/public_html/tgdclass/tgdclass.html#astro

[167] http://tgdtheory.fi/public_html/tgdlian/tgdlian.html#lianPN

[168] https://en.wikipedia.org/wiki/Dark_matter

[169] https://en.wikipedia.org/wiki/Orchestrated_objective_reduction

[170] https://en.wikipedia.org/wiki/Biophoton

[171] https://en.wikipedia.org/wiki/Quantum_critical_point

[172] https://en.wikipedia.org/wiki/Benjamin_Libet

[173] http://tgdtheory.fi/public_html/tgdeeg/tgdeeg.html#eegdark

[174] http://tgdtheory.fi/public_html/tgdeeg/tgdeeg.html#eegdark

[175]

http://tgdtheory.fi/public_html/tgdeeg/tgdeeg.html#biosupercondII

[176]

http://tgdtheory.fi/public_html/genememe/genememe.html#dnatqc

[177] Keith Conrad, "Tightening the Basic Version of Hensel's Lemma,"
        http://citeseerx.ist.psu.edu/viewdoc/download?doi=10.1.1.21
        0.7490&rep=rep1&type=pdf.

[178] Scott Carnahan, "p-adic fields for beginners,"
        http://sbseminar.wordpress.com/2007/08/21/p-adic-fields-
        for-beginners/ (21 August 2007).

ABOUT THE AUTHOR

Robert Paster offers a unique background for translating the elaborate mathematics and physics of the Topological Geometrodynamics (TGD) cognitive model to accessible Digital Mind Math.

Mr. Paster has taken advantage of his degrees in mathematics from M.I.T. and in education from Harvard to pursue careers as a teacher of mathematics, software engineer, and business mathematician, all the while keeping up with the mysteries of modern physics that have interested him since his first freshman coursework on relativity.

Robert Paster's 2006 book *New Physics and the Mind* explains the importance of TGD as a theory of quantum physics and general relativity, a theory of particle physics, cosmology, biophysics, and cognition. *Digital Mind Math* distills the essence of TGD mathematics and physics to provide an accessible explanation of how this 21st-century theory explains cognition and the mind.